EAST-WEST TECHNOLOGY TRANSFER

THE TRANSFER OF WESTERN TECHNOLOGY TO THE USSR

by
Morris Bornstein

ORGANISATION FOR ECONOMIC CO-OPERATION AND DEVELOPMENT

Pursuant to article 1 of the Convention signed in Paris on 14th December, 1960, and which came into force on 30th September, 1961, the Organisation for Economic Co-operation and Development (OECD) shall promote policies designed:

- to achieve the highest sustainable economic growth and employment and a rising standard of living in Member countries, while maintaining financial stability, and thus to contribute to the development of the world economy;
- to contribute to sound economic expansion in Member as well as non-member countries in the process of economic development; and
- to contribute to the expansion of world trade on a multilateral, non-discriminatory basis in accordance with international obligations.

The Signatories of the Convention on the OECD are Austria, Belgium, Canada, Denmark, France, the Federal Republic of Germany, Greece, Iceland, Ireland, Italy, Luxembourg, the Netherlands, Norway, Portugal, Spain, Sweden, Switzerland, Turkey, the United Kingdom and the United States. The following countries acceded subsequently to this Convention (the dates are those on which the instruments of accession were deposited): Japan (28th April, 1964), Finland (28th January, 1969), Australia (7th June, 1971) and New Zealand (29th May, 1973).

The Socialist Federal Republic of Yugoslavia takes part in certain work of the OECD (agreement of 28th October, 1961).

Publié en français sous le titre :

TRANSFERT DE TECHNOLOGIE
ENTRE L'EST ET L'OUEST
LE TRANSFERT DE TECHNOLOGIE OCCIDENTALE A L'URSS

THE OPINIONS EXPRESSED AND ARGUMENTS EMPLOYED IN THIS PUBLICATION ARE THE RESPONSIBILITY OF THE AUTHOR AND DO NOT NECESSARILY REPRESENT THOSE OF THE OECD

While self-contained and entirely the responsibility of the author, this country study was prepared within the context of a wider project concerning East-West Technology Transfer conducted under the auspices of the Committee for Scientific and Technological Policy.

It was composed of two distinct, though interrelated, phases. The first was exploratory in nature and delineated the field of inquiry. It was concluded with the publication entitled *Technology Transfer between East and West.*

The second seeks to analyse and assess two broad issues: the factors determining the assimilative capacity of the Eastern countries with regard to technology in general and Western technology in particular and the impact of Eastern imports of Western technology on East-West trade in technology.

The method of analysis consisted of country studies, analyses based on East-West and intra-CMEA trade flows, and on econometric and sectoral analyses.

Publication of this report has been authorised by the Secretary-General.

Also available

TECHNOLOGY TRANSFER BETWEEN EAST AND WEST by Eugène Zaleski and Helgard Wienert (September 1980)
(92 80 02 1) ISBN 92-64-12125-0, 436 pages £22.00 US$50.00 F200.00

EAST-WEST TECHNOLOGY TRANSFER. Study of Poland 1971-1980 by Zbigniew Fallenbuchl (September 1983)
(92 83 01 1) ISBN 92-64-12484-5, 200 pages £11.00 US$22.00 F110.00

EAST-WEST TECHNOLOGY TRANSFER
I. Contribution to Eastern Growth: An Econometric Evaluation by Stanislaw Gomulka and Alec Nove
II. Survey of Sectoral Case Studies by George D. Holliday (June 1984)
(92 84 02 1) ISBN 92-64-12565-5 96 pages £6.20 US$12.50 F62.00

EAST-WEST TECHNOLOGY TRANSFER. Study of Czechoslovakia by Friedrich Levcik and Jiri Skolka (August 1984)
(92 84 03 1) ISBN 92-64-12600-7, 102 pages £4.50 US$9.00 F45.00

* * * *

Forthcoming

EAST-WEST TECHNOLOGY TRANSFER. Study of Hungary 1968-1984 by Paul Marer

Prices charged at the OECD Publications Office.

THE OECD CATALOGUE OF PUBLICATIONS and supplements will be sent free of charge on request addressed either to OECD Publications Office, 2, rue André-Pascal, 75775 PARIS CEDEX 16, or to the OECD Sales Agent in your country.

TABLE OF CONTENTS

LIST OF TABLES

Chapter 5

Annex

THE AUTHOR

Morris BORNSTEIN is Professor of Economics at The University of Michigan (USA). His research concerns comparative economics and the Soviet and East European economies, particularly planning, pricing, economic reform and East-West economic relations. His publications include:

Comparative Economic Systems: Models and Cases, fifth edition, Irwin, Homewood, Illinois, 1985.

"Improving the Soviet Economic Mechanism", *Soviet Studies,* vol. 37, No. 1, January, 1985.

The Soviet Economy: Continuity and Change, Westview Press, Boulder, Colorado, 1981.

The author wishes to thank John Attarian for assistance in research; and Ronald Amann, John P. Hardt, John A. Martens and Heinrich Vogel for helpful comments on an earlier version of this study.

SUMMARY

Chapter 1: INTRODUCTION

This volume concerns the transfer of technology to the USSR from the developed market economies in the OECD. It examines the Soviet interest in Western technology, the modes of transfer of Western technology to the USSR, and the impact of this transfer on the Soviet economy and on Soviet foreign trade.

In this study, technology is considered knowledge necessary to apply a process, manufacture a product, or provide a service. This knowledge includes:

1. Technical information about process or product characteristics;
2. Production techniques for the transformation of labour, materials, components, and other inputs into finished outputs; and
3. Managerial systems to select, schedule, control, and market production.

Technology may be "disembodied" or "unembodied", for example in the form of technical documentation, or "embodied" in machinery or equipment. "Material transfer" of technology occurs with the import of a product by a country that cannot produce it. In the case of "design transfer", the receiver imports a plant to make the product. "Capacity transfer" entails the receiver's ability to replicate the imported plant by building additional similar facilities without foreign help. The broad conception of technology transfer includes "material transfer" because the importing country, by using the foreign machinery, equipment, or intermediate product, thereby obtains the benefits of a technology not otherwise available to it. This volume follows this comprehensive approach because it helps to explain important facets of Soviet technological development and Soviet economic relations with Western countries.

Chapter 2 of the study explains the Soviet interest in Western technology. Although rapid economic growth has long been a cardinal aim of the Soviet regime, in the last two decades new emphasis has been placed on the potential contribution of technological progress to economic growth and structural change. However, there are a number of systemic handicaps to technological progress in the USSR that, along with other factors, have caused Soviet technology to lag behind Western technology in a number of fields. As a result, the Soviet Government recognised the advantages of importing Western technology on a significant scale. But this effort has been constrained by the capacity of the USSR to assimilate foreign technology, its ability to finance convertible currency imports of technology, and (in certain fields) by export restrictions of Western governments.

The nature and scale of the various modes of transfer of Western technology to the USSR are examined in Chapter 3. First, Soviet purchases of Western licences are considered. Then the import of Western machinery and equipment is analysed. Next the role of turnkey plants is discussed. Soviet participation in other forms of industrial co-operation is also investigated. Finally, Soviet use of other methods of technology transfer, legal and illegal, is reviewed.

Chapter 4 assesses some aspects of the impact of the transfer of Western technology on the Soviet economy. This impact depends upon the nature of the technology transfer, its size in relation to Soviet economic activity, and the ability of the Soviet economy to assimilate the Western technology. Various estimates of the quantitative impact of imported Western technology are appraised. Then Soviet problems in absorbing and diffusing Western technology are examined. The analysis draws on experience reported in case studies of technology transfer in particular branches of industry.

The impact of technology transfer from the West on Soviet foreign trade is considered in Chapter 5. The size, commodity composition, and geographical distribution of Soviet foreign trade as a whole are reviewed first. Then, Soviet imports from and exports to Western developed market economies are analysed, with particular attention to how technology transfer from the West may affect subsequent Soviet exports to the West. Next, possible effects of the transfer of Western technology to the USSR on Soviet trade with Eastern Europe and with less developed market economies are discussed.

Chapter 6 summarises some major conclusions of the study and indicates some of the principal factors affecting the Soviet acquisition of Western technology in the future.

Chapter 2: SOVIET INTEREST IN WESTERN TECHNOLOGY

Technological progress has been assigned a more important role in Soviet economic growth strategy as part of the shift from an "extensive" approach based on increases in factor inputs to an "intensive" approach stressing improvement in factor productivity. But technological progress in the USSR has been hampered by systemic features of the Soviet economy, including serious and persisting problems concerning administrative organisation, financing, supply, pricing, and performance indicators and incentives. As a result, in many important fields, Soviet technology is behind that of leading Western countries. Thus, the USSR embarked on a broad programme to obtain foreign technology, involving a number of Soviet government agencies concerned with aspects of the choice, acquisition, and use of foreign technology. However, the Soviet effort to secure Western technology has been constrained by the ability of the Soviet economy to assimilate this technology, by the Soviet ability to pay for it, and, for certain kinds of technology, by export controls of some Western governments.

1. Technological Progress in the USSR

Both Soviet official statistics and Western estimates show a sharply declining trend since the early 1960s in such measures of Soviet economic performance as growth rates of national income, investment, productive fixed assets, industrial output, and industrial labour productivity. In the last 20 years most of the growth of Soviet GNP and industrial production has been due to greater factor inputs, especially of capital, while factor productivity has increased only slowly or has declined.

Because the USSR cannot raise the rates of growth of factor inputs, its efforts to bolster sagging growth rates have emphasized boosting factor productivity through technological progress – the introduction of new processes and products into production – as well as through changes in economic administration.

It might appear that in principle a socialist centrally planned economy has some advantages over capitalist market economies in the potential for technological progress. For instance, central planning agencies can initiate, organise, co-ordinate, and finance fundamental and applied research, development, and innovation on a national scale, according to official priorities, and share the results widely without restrictions of commercial secrecy. Also, central planning can enforce a high rate of investment, achieving expansion of the capital stock that can embody new process and product technologies. Further, the investment programme can be concentrated in branches of the economy with the greatest potential for technological progress – machine building rather than services, for example.

However, the effectiveness of this administrative demand from above for technological progress may be limited by other systemic features that impede technological progress by weakening the response of enterprises and research organisations to such official directives and by discouraging decentralised initiatives for innovation. In the USSR these features include, for example, problems in organisation, financing, supply, pricing, and performance indicators and incentives.

One major handicap is the organisational separation of research and development (R & D) from production in the Soviet Union. R & D are typically performed in centralised branch research facilities whose task is to develop new products and processes in response to instructions from the branch ministry, which, after accepting the results, assigns them to particular enterprises. The lack of contact between design bureaux and producing enterprises often leads the former to do work unsuited for application in production and even to be indifferent to the fate of its work once it is accepted by the ministry. Also, the introduction of R & D into production is hampered by the underfinancing of testing, pilot production, and demonstration facilities.

Enterprises may be prevented or discouraged from introducing an innovation by the chronic shortage of material inputs in the overstrained Soviet economy. Innovation has also been discouraged by Soviet pricing policy, which provides enterprises weak incentives to introduce new and cheaper processes or to produce improved models of products. In addition, the system of performance indicators for Soviet enterprises emphasizes output – more than cost, profitability in relation to capital, labour productivity, quality, introduction of new technology, savings in use of materials and fuels, etc. Further, unlike management in market economies, Soviet managers are not under competitive pressure to introduce new products and processes in order to secure profits, increase market share, and avoid business failure or takeover of the firm.

Introduction of new technology is also impeded by Soviet problems in the construction and renovation of plants. Plans for output, investment, and commissioning of capacity are poorly co-ordinated. Because of shortages of materials, equipment, and labour, projects are often completed far behind schedule.

Various attempts have been made in the last 20 years to address these problems through changes affecting planning, organisation, incentives, and contracting for research, development, and innovation (RDI). But the results of these measures have been modest, and the Soviet Government periodically repeats its call for new efforts to overcome the continuing serious problems in Soviet RDI.

These problems in the generation of technological progress in the USSR help to explain the lag of Soviet technology behind that of leading Western countries in various fields. Detailed case studies by industry specialists have shown instances of such a technology gap in iron and steel, chemicals, energy, machine tools, motor vehicles, and computers, for example. These Soviet technological lags, together with Soviet economic development priorities, explain the pattern of technology acquisitions by the USSR from the West.

2. Soviet Decision-making on the Acquisition of Western Technology

A large number of Soviet Government agencies are involved in technology acquisition decisions. They include the USSR Council of Ministers, the USSR State Planning

Committee, the USSR State Committee for Science and Technology, the USSR Academy of Sciences, the USSR State Committee for Material-Technical Supply, and the Ministries of Foreign Trade and Finance. In addition, the Communist Party Secretariat plays a leading role in overall policy on technology imports and participates in decisions on important specific acquisitions.

Soviet criteria for decisions on the acquisition of Western technology are in part inferred from the technology lags discussed above; in part explicitly indicated by leaders' policy speeches, material in plan documents and technical literature, and statements by Soviet foreign trade officials; and in part revealed by the actual pattern of Soviet technology imports from the West.

The choice of technologies to be sought from the West depends on the following:

1. Soviet plans to increase output of particular branches and products;
2. The investment – especially new plant construction – necessary to achieve the output increase;
3. The capacity of the Soviet machine building industry to supply the needed quantities of machinery and equipment at the desired technological level;
4. The ability of the USSR to obtain from East European countries what Soviet industry cannot provide;
5. The potential contribution to future Soviet exports for convertible currencies.

The application of these criteria, by the agencies mentioned, has led to substantial Soviet imports of Western technology in the motor vehicle, chemical, oil and gas, machine building, mining, and timber industries.

3. Constraints on Soviet Acquisition of Western Technology

The constraints on the acquisition of Western technology by the USSR include the Soviet ability to assimilate such technology (discussed in Chapter 4), the Soviet capability to finance technology imports, and the restrictions of Western governments on the export of technology to the USSR.

The Soviet Union's ability to finance imports of Western technology depends primarily on three related factors. One is the excess of convertible currency earnings from exports over outlays for other high-priority imports like grain. Another is the availability (and terms) of Western credits to cover deficits in the current account. The third is the extent of Soviet success in making countertrade arrangements for future exports to service the convertible currency debt.

In convertible currency transactions Soviet merchandise imports exceed merchandise exports, causing a persistent deficit in merchandise trade. In addition, net interest payments are negative. Offsetting the trade and interest deficits are arms sales to developing market economies, gold sales, and net borrowing abroad. Thus, Soviet ability to pay for Western machinery, equipment, and intermediate products like large-diameter pipe is enhanced by several kinds of developments. They include improvements in Soviet terms of trade, for instance from movements in world oil prices; higher prices on world gold markets; and greater opportunities for arms sales to Third World countries. On the other hand, Soviet purchases of Western technology tend to be curtailed by, for example, additional grain imports in years of bad Soviet harvests, and principal and interest payments to service the Soviet convertible currency debt.

The "creditworthiness" of the USSR has been rated favourably in international financial markets because the USSR has maintained at 15-17 per cent the ratio of its convertible currency debt service to the sum of its convertible currency receipts from merchandise exports, arms sales, gold sales, and other sources potentially available to make debt service payments.

The USSR endeavours through compensation agreements to link the repayment of credits for imports of Western machinery and equipment to a contracted flow of convertible currency exports from projects incorporating the imports. By industrial branch, these agreements are concentrated in natural gas, chemicals, forestry products, and metallurgy.

The Soviet Union's acquisition of certain kinds of Western technology of military or strategic significance has been constrained by unilateral or multilateral export restrictions of Western governments.

Chapter 3: MODES OF TRANSFER OF WESTERN TECHNOLOGY TO THE USSR

This chapter considers in turn licence purchases, commodity imports, turnkey plants, industrial co-operation agreements, and other modes of transfer of Western technology to the USSR.

1. Licence Purchases

Through purchases of foreign licences, the USSR acquires technology not available domestically and releases its R & D organisations for work on other products and processes. Also, in the long run licence purchases may save foreign exchange, when domestic production of a commodity replaces imports of it, or earn foreign exchange when Soviet production based on the licence is exported. However, the magnitude of Soviet licence purchases is relatively small, in comparison with license purchases of developed market economies and also in comparison with other modes of technology transfer to the USSR.

2. Commodity Imports

Commodity imports can be regarded as a mode of technology transfer when they embody technologies not available in the recipient country. Such imports transfer not the methods of production themselves, but rather the products constituting the results of these technologies. In some cases, a nation which possesses the relevant technology, but has not diffused it widely enough, may import part of its supply in order to supplement domestic production. However, not all imports can be explained by technological lags and embodied technology transfers. The level and composition of imports (and exports) also reflect inter-industry and intra-industry specialisation decisions made for comparative advantage and other reasons.

Although a technology gap may be responsible for imports of technologically advanced consumer goods, the analysis of technology transfer via commodity imports concentrates on producer goods to make other commodities. Such "embodied" technology transfer has occurred on a significant scale in Soviet imports of Western machinery, equipment, metal products, and chemicals, used primarily in the USSR's engineering and metalworking, chemical, and energy industries. (These imports are discussed further in Chapter 5.)

3. Turnkey Projects

In turnkey projects, foreign firms supply whole production systems and thus may design facilities, supervise construction and installation work, train personnel, and aid in the start-up of production. Many such turnkey projects have been carried out in the Soviet chemical, ferrous metallurgy, machine building, and light and food industries.

4. Industrial Co-operation

Through industrial co-operation agreements (ICAs), the USSR obtains the long-term involvement of Western firms in Soviet production methods, product characteristics, quality control, and training of personnel. The most common kinds of Soviet-Western ICAs are for co-production and for specialisation. In the former, the partners specialise in producing components and then exchange them so each can make the same final product. In the latter, the partners specialise in making end-products and then exchange them so each has a full line for sale. More than a third of all Soviet ICAs are in the chemical industry.

5. Other Modes

The Soviet Union has an extensive network of organisations for the collection, analysis, abstracting and translation, and dissemination of scientific and technical information in foreign publications. In addition, Soviet specialists attend many foreign industrial and trade exhibitions and scientific meetings.

Various Soviet R & D and economic organisations follow industrial trends in developed market economies through the study of Western patent data, which cover most of the important areas of industrial technology. Such patent information is used in Soviet decisions about purchases of foreign licences or goods and also in Soviet decisions about domestic R & D programmes.

Another significant mode of technology transfer is Soviet scientific exchange programmes with many Western countries. Although these programmes are arranged under broad inter-governmental agreements, the Western partners at the operating level are often national academies of sciences, cultural relations bodies, scientific associations, or universities. Detailed agreements between the Western and Soviet partners specify the particular scientific fields included, the number of "person-months" of exchange visits during a given period, and financial arrangements. Under some programmes, individual Soviet scientists participate in symposia, make short visits to Western scientific institutions, or spend as much as a year in sustained research at a particular facility. Under other programmes, joint Soviet-Western working groups present their respective research results at periodic conferences and may conduct some research jointly. Through such scientific and technical exchanges, Soviet participants acquire information about the state of knowledge and current research in the West about their fields.

The USSR has been able through several methods to obtain some of the technology covered by Western export controls. A legal method is the establishment of firms chartered as local companies in the United States or a West European country but owned by the Soviet Union (or an East European nation). Such firms may legally purchase controlled technology and study it in the country of origin, although they cannot legally export the goods or data without an export licence. Some Western technology is secured through industrial espionage. Another illegal method involves the use of intermediary firms that procure valid Western export licences for an approved end-use in an authorised country of destination and then reship the goods to the USSR.

Chapter 4: THE IMPACT OF THE TRANSFER OF WESTERN TECHNOLOGY ON THE SOVIET ECONOMY

This chapter considers:

1. Macroeconomic approaches to the estimation of the impact of the transfer of Western techology on Soviet industry;
2. Case studies of experience in particular industrial branches; and
3. Problems encountered in the absorption and diffusion phases of the assimilation of Western technology by the Soviet economy.

1. Macroeconomic Approaches

Macroeconomic estimates of the impact of Western technology on the Soviet economy are controversial because they rely on simplifying assumptions of disputed validity as well as on incomplete and imperfect statistical data. Hence, these estimates are the subject of criticism and disagreement among experts on technical grounds. The dominant view among specialists is that the aggregate impact of the transfer of Western technology on Soviet industry cannot really be measured satisfactorily but is likely to be rather modest (and the impact on the Soviet economy as a whole, of which industry is only a part, even more so).

Western machinery and equipment are a minor element in total Soviet investment and capital stock. It is reasonable to expect that because of their technological superiority Western machinery and equipment could have a disproportionately large effect in expanding Soviet industrial output. But it is not possible to measure this effect satisfactorily with the available statistical data and the current state of the art in econometrics. The most that expert opinion can offer on the subject is a tentative, speculative, order-of-magnitude conjecture that imports of Western machinery and equipment did not contribute more (and may have contributed less) than half a percentage point of the annual growth of Soviet net industrial output in the 1970s.

2. Case Studies

Although the impact of technology transfer from the West on the overall growth of Soviet industry or the Soviet economy is relatively small, this impact might be more significant for particular branches of industry, products, or projects. Case studies provide information on the nature and size of the effects of technology transfers from the West, on how these effects were achieved, and on the factors limiting them.

Chemicals. The dominant mode for the transfer of Western technology to the Soviet chemical industry has been large-scale purchases of machinery and equipment, often for turnkey projects and frequently complemented by industrial co-operation agreements. The USSR turned to the West for chemical technology and equipment partly because of Soviet weaknesses in the research and development stages for many chemical products and processes, partly because of problems in carrying promising development through to large-scale production, and partly because of the inability of Soviet metallurgy and metalworking to furnish adequate tanks and tubes for the chemical industry. The bulk of Western chemical equipment bought by the USSR has been for the production of fertilizers (ammonia and urea) and artificial fibres (polyester fibre, polypropylene, cellulose triacetate, and intermediates for making nylon and other synthetic fibres).

Motor vehicles. Western technology played an important role in the development of Soviet production of both passenger cars and trucks. Existing Soviet automotive plants were producing insufficient quantities of vehicles technologically obsolete in engineering, performance, and quality. The USSR obtained Western designs, machine tools, and production know-how from the French firm Renault to expand the production of Moskvich cars and from the Italian firm FIAT to produce a new car called the "Zhiguli" in the USSR and the "Lada" abroad. Large amounts of Western machinery and equipment were used also in the large new Kama Automotive Plant, which produces diesel engines and trucks.

Machine tools. The USSR has obtained Western machine tool technology through the study of Western published literature, acquisition of Western licences for design and production, purchases of Western machine tools, and industrial co-operation agreements (ICAs). Grinding machines, automatic lathes, and numerically-controlled machine tools of all types dominate Soviet imports of Western machine tools. Through ICAs with Western firms the USSR is obtaining design technology for conventional and advanced machine tools and production technology for advanced types.

Energy. Technology transfer from the West to the USSR has been modest in regard to coal, nuclear energy, and electric power, but more significant for oil and gas. The Soviet Union has imported from the West drill pipe that permits deeper drilling than Soviet pipe, and drilling rigs that make possible faster, deeper, and wider drilling than Soviet rigs. Also, the Soviets have obtained offshore drilling, exploration, and production equipment from the West, for use in Baltic and Arctic waters and off Sakhalin Island. The USSR has bought complete oil refineries from the West, although the technologies were generally not advanced except for the use of minicomputers and microprocessors in control systems. Western large-diameter pipe and compressors have made a substantial contribution to Soviet gas pipeline transport (and Soviet gas exports to Western Europe) by reducing construction and installation costs and by increasing operating productivity of Soviet pipelines.

Forest products. The Soviet Union has imported Western machinery and equipment for the paper, pulp, and wood processing industry. Turnkey projects and other forms of industrial co-operation – often involving compensation agreements – have been important methods of technology transfer in this industry. Through the use of Western technology the USSR seeks to equip large-scale enterprises in severe climatic areas with labour-saving machinery, to achieve more efficient processing of wood into wood products, and to raise convertible currency earnings from exports of forest products.

The technology transfer from the West to the USSR examined in these case studies raised the technological level of the particular branches of Soviet industry. But technological progress in these activities has continued in the developed market economies. Hence, the specialised technical and economic literature does not indicate any significant narrowing of the technological lag of the USSR behind the West in these fields.

3. Soviet Assimilation of Western Technology

The assimilation process has two main phases. The absorption phase involves the successful exploitation of the Western technology in the first facility for which it is acquired. The diffusion phase entails the replication of the Western technology in other plants. Weaknesses in either phase reduce the impact from the acquisition of Western technology and can thus influence Soviet decisions about the scale, modes, and fields of subsequent technology imports from the West.

No comprehensive or systematic evaluations of Soviet experience in the absorption of Western technology have been published in the Soviet general or specialised press. However, Soviet publications reveal various instances in which first the completion of facilities with Western machinery and equipment and then the operation of these facilities were hampered by poor planning and co-ordination, reflected in shortages of labour, materials and components, fuel and power, and transport.

Some information about Soviet experience in the absorption of Western technology is available from surveys of Western firms that provide machinery and equipment, including turnkey plants, and technical assistance to the USSR. Significant delays commonly occurred between the delivery of equipment by these firms and its installation and use in production. The main reasons were shortages of construction labour, and poor qualifications and motivation of available personnel; shortages of transportation and of complementary inputs from Soviet industry, like high-grade steel and electronic components; and lack of relevant manufacturing know-how.

In some cases, survey respondents believed that the Soviet plants, once commissioned, were operating below the levels that would have been attained in the West. The reasons included non-completion of related plants, shortages of materials and transport, use of unsuitable materials, failure to keep working areas sufficiently free of dirt and dust, and lack of personnel qualified to operate highly automated plants. However, the labour force in these plants was sometimes much larger than in comparable West European facilities.

Diffusion may be viewed as the final stage in the process of technology transfer, as it concerns the spread of new (in this case imported Western) technology from its initial application to other units in the economy. The speed and extent of diffusion show the economy's ability to replicate (and even to improve) the original technology and to use it for production in large volume at satisfactory cost and quality. The rate of diffusion is commonly measured by the share of total production capacity, output, or perhaps employment attributable to the new product or process.

However, the published Soviet literature offers little information of this sort about the technology imported from the West since 1970. Western firms supplying (embodied and disembodied) technology to the USSR generally know less about the diffusion phase than about the absorption phase of Soviet assimilation experience. Furthermore, because absorption and diffusion of a new Western product or process may easily take 5-10 years, evidence of successful diffusion of some Western technology imported in the mid-1970s would not be available until some time in the mid-1980s.

It appears likely that diffusion of Western technology in the Soviet economy would be difficult for a number of reasons. First, Soviet investment in fixed capital is growing slowly. Also, Soviet innovation tends to be confined to only a part of total investment in an industrial branch (like chemicals or iron and steel), with major shares of new investment still embodying traditional technologies. In contrast, in Western developed market economies new capacity is devoted entirely to new technologies. Furthermore, in the USSR diffusion is reduced by the Soviet practices of low depreciation rates and the tardy withdrawal of obsolete facilities from production.

The centrally planned economy of the USSR lacks the domestic and foreign competitive pressures that stimulate the diffusion of new technology in market economies. Also in the USSR the diffusion process is hampered by the shortcomings in organisation, financing, pricing, performance indicators, and incentives that retard Soviet technological progress generally. Shortages of Soviet labour, materials, and equipment delay the construction of new facilities and curtail their output once they are completed. In many cases, diffusion would still require at least some Western machinery and equipment, entailing convertible currency expenditures. Finally, Soviet national security restrictions may hamper the diffusion of new technology from military to civilian uses.

Despite these assimilation problems, it is reasonable to expect that the USSR will – albeit slowly and imperfectly – absorb and then diffuse part, if not all, of the technology imported from the West. However, the impact on Soviet industry and the Soviet economy as a whole will be modest, although noticeable in some particular sub-branches, such as passenger cars and certain chemicals.

Chapter 5: THE IMPACT OF THE TRANSFER OF WESTERN TECHNOLOGY ON SOVIET FOREIGN TRADE

An appraisal of the impact of the transfer of Western technology on Soviet foreign trade requires an examination of both the commodity composition of Soviet foreign trade and its geographical distribution by major market areas.

1. Aggregate Soviet Foreign Trade

According to Soviet official statistics, machinery and equipment represent about one-third, and food products another one-fourth, of the USSR's total imports. Petroleum and petroleum products constitute more than one-third, and machinery and equipment about an eighth, of total Soviet exports. About two-fifths of the USSR's total foreign-trade turnover (imports plus exports) is with the six East European centrally planned economies. Developed market economies account for about one-third, and less developed market economies about one-eighth, of total Soviet foreign-trade turnover.

However, Soviet official foreign-trade statistics in "foreign-trade rubles" combine:

1. Soviet trade with developed and less developed non-Communist countries at world market prices (and chiefly in convertible currencies); and
2. Soviet trade with Communist countries at Council for Mutual Economic Assistance "contract" prices (and primarily through accounts in "transferable rubles" that are inconvertible and non-transferable).

The validity of this aggregation, and thus of calculations of shares of components in totals – though common in Soviet publications and also in many Western studies – is questionable. Hence, it is important to consider separately Soviet trade with different market areas involving different prices and different kinds of currency.

2. Soviet Trade with OECD Countries

Manufactured goods – especially engineering products and iron and steel products – dominate Soviet imports from developed market economy countries. However, the share of food products, principally cereals, has risen sharply since 1970 as a result of heavy Soviet grain purchases in response to poor Soviet harvests. Food now accounts for about one-fourth of total Soviet imports from OECD countries.

By *branch of origin,* machinery, equipment, and metal products; metallurgy; and chemicals dominate *Soviet imports of industrial goods* from OECD countries. Within the category of machinery, equipment, and metal products, the most important sub-branches are metal- and wood-working machinery, ships and motor vehicles, construction and mining machinery, and scientific, measuring, and control equipment.

In regard to *Soviet imports* from OECD countries *of capital goods by type of product,* the largest and fastest-growing category (one-third of the total in 1982) is commodities for liquid fuel, gas, and water distribution – mainly tubes and pipes. Next in importance are ships, construction and mining machinery, and machine tools.

Analysis of the distribution of the total value of *Soviet imports of capital goods by end use* shows that in 1982 about one-half went to Soviet industry and about one-fourth to oil and gas distribution. Of the amount for industry in 1982, engineering and metalworking received about one-sixth, and mining and fuel extraction machinery another one-sixth.

In *Soviet imports* from OECD countries *of technology-based intermediate goods,* the most important categories are iron and steel, chemicals, and plastics.

In short, Soviet imports of manufactured goods from the OECD countries have been concentrated in machinery, equipment, metal products, and chemicals, and they have been destined primarily for the Soviet engineering and metalworking, chemical, and energy industries.

The share of "technology-based" products in total exports of OECD countries to the USSR was 57 per cent in 1982 (below a high of 73 per cent in 1970 and 1974). The rest of OECD countries' exports to the USSR, 43 per cent of the total in 1982, consisted of products judged not to be "technology-based". The combined share of goods of "high" and "moderate" R & D intensity in the OECD countries' total exports to the USSR ranged from 7 to 14 per cent during 1970-1982.

Primary products have accounted for over four-fifths of total Soviet exports to OECD countries. Fuels alone constituted more than three-fourths of the total in 1982, compared with one-third in 1970.

"Technology-based" goods represented only 9 per cent of total Soviet exports to OECD countries in 1982 (down from 23 per cent in 1970 and 27 per cent in 1973).

The link between the transfer of Western technology to the USSR and subsequent Soviet exports to the West is clear in the case of natural gas. Almost all Soviet exports of natural gas to Western Europe are under compensation agreements through which the USSR earlier obtained from developed market economies large-diameter pipe and compressor and other equipment for pipelines from Soviet gasfields to Eastern and Western Europe.

The Soviet chemical industry has made large purchases of Western machinery and equipment, often for turnkey projects and frequently complemented by industrial co-operation agreements. The bulk of the Western technology was for the production of fertilizers and artificial fibres. The subsequent growth of Soviet chemical production was reflected in rising exports to the OECD countries, particularly of ammonia, but also of other products like carbamide, polyethylene, and acrylonitrile. Further increases in Soviet chemical exports to the West are scheduled under compensation and barter agreements.

The USSR exports under the name "Lada" a version of the passenger car modelled on the FIAT-124 that is sold in the USSR as the "Zhiguli". It also exports the "Niva", a four-wheel-drive vehicle of the jeep or land-rover type that is of Soviet conception but based on FIAT technology. Both the Lada and the Niva are sold in Western markets at about 75 per cent of the price of similar West European and Japanese vehicles. This differential reflects in part the Lada's now somewhat outdated design and in part aggressive Soviet pricing. The Soviet share of the Western car market is insignificant.

The Soviet Union has also obtained Western technology for machine tools, especially numerically-controlled tools, and as a result it now exports numerically-controlled and other machine tools incorporating Western technology. Because these Soviet tools are technologically less advanced than current Western models, Soviet sales of these tools to the West are small.

On the whole, there is little evidence of a strong competitive threat in Western markets by Soviet exports attributable to the transfer of Western technology to the USSR.

3. Soviet Trade with Eastern Europe

Fuels constitute half of Soviet exports to Eastern Europe, and machinery and equipment only a sixth. Machinery and equipment account for three-fifths of total Soviet imports from Eastern Europe, and other manufactured goods an additional one-fifth. Soviet imports of capital goods from Eastern Europe are widely distributed across the Soviet economy. The end-user branches with the largest shares, agriculture and electricity, each receive less than one-tenth of total imports of capital goods from Eastern Europe.

To some extent, Soviet imports of machinery and equipment from Eastern Europe may be explained by the technological superiority of East European products compared to Soviet products. But another important factor is Council for Mutual Economic Assistance specialisation and co-operation agreements, among producers at a similar technological level, intended to achieve economies of scale. However, on the whole the technological level of Soviet machinery and equipment imports from Eastern Europe is below that of machinery and equipment imported by the Soviet Union from the West.

There is little evidence of technology transfer from the West to the USSR through Eastern Europe, rather than directly.

In some cases, Soviet exports to Eastern Europe can be attributed to earlier Soviet acquisition of Western technology. One example is the sale of natural gas to Eastern Europe through pipelines built with Western pipe and compressor station equipment. Another instance is exports to Eastern Europe of cars and trucks from Soviet plants that incorporate Western technology.

4. Soviet Trade with Less Developed Market Economies

Food, crude materials, and mineral fuels account for about 90 per cent of Soviet imports from less developed market economies. The main Soviet exports to these countries are arms, machinery and transport equipment, and mineral fuels.

There is no significant indication that less developed market economies transfer to the USSR technology they have acquired from the West, or that the USSR sells to these nations technology it previously obtained from the West.

Chapter 6: CONCLUSION

The USSR will continue to import Western technology through a variety of transfer modes for many branches of the Soviet economy. Average annual growth rates in the USSR in 1984-1988 are likely to be less than 0.1 per cent for the labour force and in the 3-4 per cent range for investment. Hence, the Soviet regime will strive to achieve improvements in factor productivity, in which technological progress plays a key role. However, the systemic handicaps to technological progress in the USSR are serious. There is no evidence of significant success in overcoming them through the various measures taken to improve the research, development, and innovation process. Nor is technological progress – or economic growth generally – likely to be increased much by Soviet efforts at economic system reforms undertaken since 1979. Thus, the Soviet need for Western technology will remain strong.

But the scope and content of future Soviet acquisition of Western technology will depend upon a number of factors. They include the nature of the Soviet investment programme, the assessment of the USSR's experience in the acquisition and assimilation of Western technology, financing constraints, and Western governments' export restrictions.

The Soviet leadership faces various policy issues affecting the acquisition of Western technology:

1. What will be the size and composition of the investment programme?
2. To what degree must Western technology surpass Soviet (or East European) technology in order to justify the choice of Western technology, despite its convertible currency cost and its vulnerability to possible Western export restrictions?
3. To what extent should technology acquisition policy stress "design transfer" rather than "material transfer"?
4. How much convertible currency should the USSR try to mobilise for the acquisition of Western technology – for instance, by redirecting oil exports from Eastern Europe to Western Europe, drawing on gold reserves, restraining grain purchases, and/or increasing the convertible currency debt-service ratio?

In turn, there are policy issues for Western governments concerning the transfer of technology to the USSR:

1. To what degree and in what ways should unilateral or multilateral export restrictions of Western governments be increased or reduced?
2. What will be the role of Soviet oil and gas in the energy supply of Western Europe?
3. In what amounts and on what terms should Western countries provide official or private credit to the Soviet Union?

In view of the Soviet need for Western technology, on the one hand, and the constraints on its acquisition, on the other, it is reasonable to expect that Soviet imports of Western technology will continue, but on a more modest scale and a more selective basis than in the past decade. Proposed acquisitions of Western technology will be evaluated according to stricter standards for the priority of different industries (branches, product groups), the superiority of Western technology over Soviet (or East European) alternatives, the probable ease of assimilation of Western technology, and its potential contribution to increasing convertible currency exports or reducing convertible currency imports.

Chapter 1

INTRODUCTION

This volume concerns the transfer of technology to the USSR from the developed market economies in the OECD. Technology involves certain kinds of knowledge relevant to production. Various aspects of the nature and results of the transfer of Western technology to the USSR are analysed and evaluated in this study. It examines the Soviet interest in Western technology, the modes of transfer of Western technology to the USSR, and the impact of this transfer on the Soviet economy and on Soviet foreign trade.

Technology is knowledge necessary to apply a process, manufacture a product, or provide a service. This knowledge includes:

1. Technical information about process or product characteristics;
2. Production techniques for the transformation of labour, materials, components, and other inputs into finished outputs; and
3. Managerial systems to select, schedule, control, and market production.

Technology may be "disembodied" or "unembodied", for example in the form of technical documentation, or "embodied" in machinery or equipment.

In turn, the transfer of technology between organisations and between nations has several dimensions, including:

1. Its content;
2. Whether payment is involved;
3. The extent of negotiation;
4. Modes of transfer; and
5. The degree of success[1].

1. There are differences in the content of the technology transferred. "Material transfer" occurs with the import of a product by a country that cannot produce it. In the case of "design transfer", the receiver imports a plant to make the product. Finally, "capacity transfer" entails the receiver's ability to replicate the imported plant by building additional similar facilities without foreign help. Some analysts prefer to consider as technology transfer only the second and third types, excluding the first on the ground that technology transfer concerns acquisition of the ability to make a product, rather than only to use it. In contrast, the broader conception of technology transfer includes the first type on the ground that the importing country, by using the foreign machinery, equipment, or intermediate product, thereby obtains the benefits of a technology not otherwise available to it. The broader view is more common in both Western and Soviet literature on technology transfer from the West to the USSR. The present volume adopts this more comprehensive approach because it helps to explain important facets of Soviet technological development and Soviet economic relations with Western countries.

2. "Commercial" transfers, like the purchase of machinery, involve a payment by the receiver to the supplier. In contrast, "non-commercial" transfers, for instance through the study of technical literature, do not.

3. The transfer of technology may or may not be consciously negotiated between the supplier and the receiver. In "negotiated" transfers, the two parties contract for shipments of machinery on a significant scale, licences, and/or scientific-technical and other kinds of industrial co-operation. In the transfer of Western technology to the USSR, such negotiations by private Western firms are sometimes subject to regulation by Western governments through export controls, official credit support, and less formal suasion. In contrast, the "unnegotiated" transfer of technology without the conscious co-operation of the supplier can occur through, for instance, review of published technical literature, small legal purchases of individual machines for reverse engineering (analysis of its technical features intended to enable the receiver to reproduce the machine), and industrial espionage.

4. There are various modes (channels) of international technology transfer. *a)* Foreign direct investment by multinational corporations (MNCs) in subsidiaries abroad is a prominent mode of technology transfer among developed market economies and from them to less developed market economies. The MNC typically furnishes capital and management as well as technology. In the case of a joint venture, the firm in the recipient country is partly owned by local private or government organisations. Foreign direct investment is commonly considered an especially effective method of technology transfer because it facilitates long-term, frequent, and iterative flows of firm-specific as well as more generally available technology. The exchange and training of technical and managerial personnel play an important role in this regard. However, the USSR does not permit foreign direct investment in its territory. The limited Soviet participation in joint ventures with Western firms occurs in Western countries or in third countries in less developed regions.

Thus, an important mode of transfer of Western technology to the USSR is *b)* the sale of machinery and equipment, including complete "turnkey" plants. *c)* Some technology has been transferred through licences to use patents, production know-how, or trademarks. *d)* Technology transfer also occurs through a variety of industrial co-operation arrangements, like co-production, sub-contracting, and scientific-technical agreements. *e)* In addition, there are many less formal modes of technology transfer, such as the flow of publications, personal contacts (including training abroad and participation in technical conferences), and illegal methods of acquiring technology.

5. Technology transfer may be more or less successful, depending upon the recipient's ability to assimilate – i.e., first absorb and then diffuse – the foreign technology.

This study provides a concise analysis and evaluation of selected key aspects of technology transfer from the West to the USSR since 1970, drawing upon and integrating information from Soviet as well as Western sources.

Chapter 2 explains the Soviet interest in Western technology. Although rapid economic growth has long been a cardinal aim of the Soviet regime, in the last two decades new emphasis has been placed on the potential contribution of technological progress to economic growth and structural change. However, there are a number of systemic handicaps to technological progress in the USSR that, along with other factors, have caused Soviet technology to lag behind Western technology in a number of fields. As a result, the Soviet Government recognised over a decade ago the advantages of importing Western technology on a significant scale. But this effort has been constrained by the capacity of the USSR to assimilate foreign technology, its ability to finance convertible currency imports of technology, and (in certain fields) by Western governments' export restrictions.

The nature and scale of the various modes of transfer of Western technology to the USSR are examined in Chapter 3. First, Soviet purchases of Western licences are considered. Then, the import of Western machinery and equipment is analysed in detail. Next, the role of turnkey plants in the Soviet technology transfer programme is discussed. Soviet participation in other forms of industrial co-operation is also investigated. Finally, Soviet use of other methods of technology transfer, legal and illegal, is reviewed.

Chapter 4 assesses some aspects of the impact of the transfer of Western technology on the Soviet economy. This impact depends upon the nature of the technology transfer, its size in relation to Soviet economic activity, and the ability of the Soviet economy to assimilate the Western technology. Various estimates of the quantitative impact of imported Western technology are summarised and appraised. Then Soviet problems in absorbing and diffusing Western technology are examined. The analysis draws on experience reported in case studies of technology transfer in particular branches of industry.

The impact of technology transfer from the West on Soviet foreign trade is considered in Chapter 5. First, the size, commodity composition, and geographical distribution of Soviet foreign trade as a whole are reviewed. Then, Soviet imports from and exports to Western developed market economies are analysed, with particular attention to how technology transfer from the West may affect subsequent Soviet exports to the West through countertrade and other ways. Next, Soviet trade with Eastern Europe is examined to assess possible effects of the transfer of Western technology to the USSR. Finally, implications of this technology transfer for Soviet trade with less developed ("Third World") countries are discussed.

Chapter 6 summarises some major conclusions of the study and indicates some of the principal factors affecting the future evolution of Soviet acquisition of Western technology.

With the scope just described, this study does not focus on the participation of individual Western countries in technology transfer to the USSR, on the impact of this transfer on Western economies, or on the technology transfer policies of Western governments[2]. Nor does the study deal with weapons technology, since a thorough analysis of this subject would involve classified information not available to the author or appropriate for this study.

NOTES AND REFERENCES

1. For more detailed discussion of technology transfer concepts and alternative terminologies, see, for example, Philip Hanson, *Trade and Technology in Soviet-Western Relations* (New York: Columbia University Press, 1981), pp. 6-15, and George Holliday, *Technology Transfer to the USSR, 1928-1937 and 1966-1975: The Role of Western Technology in Soviet Economic Development* (Boulder, Colo.: Westview Press, 1979), pp. 10-23.
2. These and other topics, such as comparisons of Soviet and East European technology transfer experience, are covered in the more comprehensive volume by Eugene Zaleski and Helgard Wienert, *Technology Transfer between East and West* (Paris: Organisation for Economic Co-operation and Development, 1980). Another relevant important recent book is Hanson, *Trade*.

Chapter 2

SOVIET INTEREST IN WESTERN TECHNOLOGY

The Soviet Union has imported Western technology in various fields in an effort to stimulate technological progress in order to promote the growth and modernisation of the economy. However, the Soviet Union's ability to acquire Western technology has been limited by various factors that influence the nature, modes, and scale of technology transfer from the West. Section A of this chapter analyses reasons for the Soviet pursuit of Western technology. Next, Section B examines Soviet decision-making on technology imports from the West, including the administrative apparatus and the choice criteria. Then, Section C considers constraints on Soviet imports of Western technology, arising from Soviet assimilative capacity, Soviet financing ability, and Western government policies.

Technological progress has been assigned a more important role in Soviet economic growth strategy as part of the shift from an "extensive" approach based on increases in factor inputs to an "intensive" approach stressing improvement in factor productivity. But technological progress in the USSR has been hampered by systemic features of the Soviet economy, including serious and persisting problems concerning administrative organisation, financing, supply, pricing, and performance indicators and incentives. As a result, in many important fields, Soviet technology is behind that of leading Western countries, according to detailed expert studies of individual industries and product groups. Thus, the USSR embarked on a broad programme to obtain foreign technology, involving a number of Soviet Government agencies concerned with aspects of the choice, acquisition, and use of foreign technology. However, the Soviet effort to secure Western technology has been constrained by the ability of the Soviet economy to assimilate this technology, by the Soviet ability to pay for it, and, for certain kinds of technology, by export controls of some Western governments.

A. TECHNOLOGICAL PROGRESS IN THE USSR

A comprehensive discussion of the many complex facets of Soviet technological progress is beyond the scope of this study[1]. This section addresses only three related aspects of technological progress in the USSR that are closely linked to Soviet decisions on the import of Western technology. First, the role of technological progress in Soviet economic growth is considered. Then, some systemic handicaps to technological progress in the USSR are discussed. Finally, the resulting lag, or gap, between Soviet and Western technology in various parts of industry is examined.

1. Technological Progress in Soviet Economic Growth

The long-standing desire of the Soviet regime for rapid economic growth – emphasising a high rate of investment and the development of heavy industry – as a basis for economic and military power is well known[2].

Soviet leaders were therefore seriously concerned by the slowdown in economic growth that began in the mid-1950s, according to published official Soviet statistics. Table 1 presents official data on rates of growth of national income and of selected other indicators of economic activity closely linked with technological progress, such as productive fixed assets, industrial production, and industrial labour productivity. The table shows the sharply (though not always monotonically) declining trend in all of these measures of Soviet economic performance.

Table 1. **Selected official measures of Soviet economic growth, selected periods, 1951-1983**

Per cent

	National Income Produced[a]	National Income Utilised[b]	Productive Fixed Assets[c]	Gross Industrial Production	Industrial Labour Productivity
Average annual rate of growth[d]					
1951-55	11.4	*e*	10.3	13.0	8.2
1956-60	9.1	*e*	9.5	10.4	6.5
1961-65	6.5	5.9	9.7	8.6	4.6
1966-70	7.7	7.1	8.1	8.5	5.7
1971-75	5.7	5.1	8.7	7.4	6.0
1976-80	4.2	3.9	7.4	4.4	3.2
Rate of growth over preceding year					
1976	5.9	5.3	7.8	4.8	3.0
1977	4.5	3.5	7.4	5.7	3.9
1978	5.1	4.5	7.7	4.8	3.7
1979	2.2	2.0	7.0	3.4	2.7
1980	3.5	3.8	7.0	3.6	2.6
1981	3.2	3.2	6.9	3.4	2.7
1982	3.4	2.6	6.7	2.8	2.1
1983	*e*	3.1	*e*	4.0	3.5

a) Net material product.
b) Differs from "national income produced" because of foreign trade balance and various "losses".
c) Including livestock but excluding fixed assets in housing, health, education, and other "non-productive" branches.
d) The base year is the year before the stated period.
e) Not available.
Sources: USSR, Tsentral'noe statisticheskoe upravlenie, *Narodnoe khoziaistvo SSSR v 1965 godu* [National economy of the USSR in 1965] (Moscow: Statistika, 1966), pp. 57-62, and *Narodnoe khoziaistvo SSSR v 1980 godu* [National economy of the USSR in 1980] (Moscow: Statistika, 1981), pp. 39-46 ; *Narodnoe khoziaistvo SSSR v 1982 godu* [National economy of the USSR in 1982] (Moscow: Statistika, 1983) pp. 40-48 ; *Pravda*, 24th January 1982, pp. 1-2, 23rd January, 1983, pp.1-2, and 29th January, 1984, pp.1, 3.

These (and other) official Soviet statistics are important because they serve as the basis for Soviet assessments of economic performance and decisions about responses. However, these official statistics have various conceptual and methodological limitations. For example, the Soviet concept of "national income" refers to net material product, excluding most services, and the industrial production series covers the total (gross) output of all reporting units, without excluding purchases of intermediate products by one industrial unit from another[3].

Hence, Western analyses of Soviet economic activity usually use (in addition or instead) the broader concept of gross national product (GNP) rather than net material product, and net rather than gross industrial production. Table 2 presents a widely accepted set of Western estimates of rates of growth of Soviet GNP and its components. Table 3 contains estimates of rates of growth of net industrial production (excluding intra-branch purchases) by branch of industry. Although the coverage of Tables 2 and 3 thus differs in significant ways from that of Table 1, these Western estimates confirm the conclusion of a sharp deceleration of Soviet economic growth.

Table 2. **Rates of growth of Soviet gross national product at factor cost, by end use, selected periods, 1961-1982[a]**

Per cent

	Average Annual Rate of Growth[b]								
	1961-65	1966-70	1971-75	1976-80	1978	1979	1980	1981	1982[c]
Gross National Product	5.0	5.3	3.7	2.7	3.4	0.4	1.7	2.2	2.0
Consumption[d]	3.7	5.3	3.5	2.8	3.0	2.7	3.1	2.1	1.5
Investment	7.6	6.0	5.4	4.1	3.7	1.5	2.5	3.7	2.6
New fixed[e]	7.2	6.4	4.8	3.7	3.1	1.2	2.1	3.3	2.2
Defence, administration, R&D inventory change, net exports, and outlays not elsewhere classified	5.7	3.9	1.4	–0.9	4.1	–9.7	–5.5	–1.2	3.0

a) 1970 end-use weights have been revised from an established price basis to a factor cost basis by the subtraction of turnover taxes and profits and the addition of implicit amortization and capital charges and subsidies. These revised weights are then moved over time by indexes of the various end uses in constant prices.
b) The base year is the year before the stated period.
c) Preliminary estimate.
d) Including consumption of food, soft goods, durables, personal services, and personal and government outlays for education and health.
e) Including machinery and equipment, construction and other capital outlays, and net additions to livestock.
Source: US Central Intelligence Agency, *Handbook of Economic Statistics 1983* (CPAS 83-10006; Washington, D.C., September 1983), Table 39, p. 62.

Table 3. **Rates of growth of Soviet industrial, production, by branch of industry, selected periods, 1961-1982[a]**

Per cent

	Average Annual Rate of Growth[b]								
	1961-65	1966-70	1971-75	1976-80	1978	1979	1980	1981	1982[c]
Total industrial production	6.5	6.3	5.9	3.2	3.3	2.1	2.8	2.5	2.2
Industrial materials	6.8	5.8	5.4	2.4	2.8	0.2	2.2	1.9	1.1
Electricity	11.5	7.9	7.0	4.5	4.7	2.9	4.5	2.5	3.0
Fuels	6.3	5.0	5.0	3.1	3.2	2.9	1.8	1.5	2.1
Ferrous metals	7.2	5.1	4.0	1.0	2.2	*d*	–0.5	–0.1	–0.9
Non-ferrous metals	7.7	7.5	5.9	2.3	2.1	2.4	1.4	1.3	0.8
Wood, pulp, and paper products	2.6	2.9	2.6	–0.3	–0.4	–2.9	1.7	2.3	1.4
Construction materials	5.4	5.8	5.4	1.2	3.3	–4.6	0.5	1.4	–1.4
Chemicals	12.0	8.9	8.6	3.6	3.6	–0.2	4.7	4.0	1.6
Machinery	7.2	7.0	8.0	5.0	5.0	4.3	4.1	3.4	3.8
Producer durables	8.6	7.9	8.9	5.7	6.0	5.4	4.5	3.1	4.2
Consumer durables	9.6	11.3	11.7	6.8	8.4	4.5	7.6	7.2	1.1
Non-durable consumer goods	4.9	6.4	3.4	1.8	0.6	2.5	1.4	1.9	1.5
Soft goods	2.6	7.2	2.7	2.7	2.6	1.8	2.3	1.9	–0.1
Processed foods	6.9	5.8	3.9	1.1	–1.1	3.1	0.7	1.9	2.8

a) Based on US Central Intelligence Agency estimates rather than Soviet official series. The latter are believed to contain an upward bias in rates of growth because of double-counting and disguised inflation. The branch indexes shown above are formed by combining a sample of products in which intrabranch purchases have been excluded. The indexes for industrial materials, consumer non-durables, and total industrial production are formed by combining the component branch indexes with 1970 value-added weights.
b) The base year is the year before the stated period. Growth rates are derived from unrounded data.
c) Preliminary estimate.
d) Negligible.
Source: US Central Intelligence Agency, *Handbook of Economic Statistics 1983* (CPAS 83-10006; Washington, D.C., September 1983), Table 46, p. 67.

A priori, one might expect that Soviet economic growth would inevitably slow down as the economy became larger – a common phenomenon in the historical experience of the more developed countries. A standard method of quantitative analysis of the causes of changes in rates of economic growth is through production functions that attribute the growth of output to the growth of factor inputs, on the one hand, and the growth of factor productivity, on the other. The best known and most widely used approach is a Cobb-Douglas production function which assigns weights to each of the factors (labour, capital, and land – or sometimes only labour and capital) intended to measure their marginal contribution to output and totalling unity. The growth in factor productivity is then obtained as a "residual" – the difference between the rate of growth of output and the weighted sum of factor input increases. Such Cobb-Douglas estimates are presented in Tables 4 and 5 because they are conceptually familiar to economists, intuitively understood by non-specialists, and available in consistently computed series for long time periods.

Table 4 shows the roles of factor input changes and factor productivity changes in the growth of Soviet GNP since 1961. Most of the GNP growth has been due to increased factor inputs, especially of capital. Combined factor productivity has grown only slowly or has declined during the period, as a result of the decrease in the marginal productivity of capital.

In Table 5 a similar analysis is applied to industrial production, although only labour and capital are counted as inputs, since the contribution of land to industrial production is considered negligible. Here also output increases are chiefly explained by increases in inputs, notably capital, while factor productivity grows modestly or declines as a result of the fall in the marginal productivity of capital.

These important findings indicated by Tables 4 and 5 are widely accepted by Western specialists on the Soviet economy. However, there are many other production function analyses of Soviet economic growth – prepared with different objectives, concepts, methodologies, assumptions, and data bases[4]. A review of these analyses is outside the scope of this

Table 4. **Rates of growth of Soviet aggregate factor productivity, selected periods, 1961-1982**

Per cent

	Average Annual Rate of Growth[a]								
	1961-65	1966-70	1971-75	1976-80	1978	1979	1980	1981	1982
Gross national product[b]	5.0	5.3	3.7	2.7	3.4	0.4	1.7	2.2	2.0
Inputs									
Man-hours	1.6	2.0	1.7	1.3	1.7	1.2	1.1	1.0	1.1
Capital	8.8	7.4	8.0	6.9	6.8	7.0	6.5	6.7	6.1
Land	0.6	-0.3	0.8	c	c	c	c	c	c
Man-hours, capital, and land combined[d]	4.5	4.1	4.2	3.6	3.7	3.5	3.2	3.2	3.1
Factor productivity									
Man-hours	3.4	3.2	2.0	1.3	1.7	-0.8	0.6	1.2	0.9
Capital	-3.5	-2.0	-4.0	-4.0	-3.2	-6.1	-4.6	-4.2	-3.8
Land	4.4	5.6	2.9	2.7	3.4	0.4	1.4	2.2	2.0
Man-hours, capital, and land combined	0.5	1.1	-0.5	-0.8	-0.3	-3.0	-1.5	-1.0	-1.1

a) The base year is the year before the stated period.
b) Based on indexes of gross national product in 1970 rubles by sector of origin at factor cost.
c) Negligible.
d) Inputs of man-hours, capital, and land are combined with weights of 55.8 per cent, 41.2 per cent, and 3.0 per cent, respectively, in a Cobb-Douglas (linear homogeneous) production function. These weights represent the distribution of labour costs (wages, other income, and social insurance deductions), capital costs (depreciation and a 20-per cent charge on gross fixed capital, including livestock), and land rent in 1970, the base year for all indexes underlying the growth rate calculations.

Source: US Central Intelligence Agency, *Handbook of Economic Statistics 1983* (CPAS 83-10006; Washington, D.C., September 1983), Table 44, p. 66.

Table 5. **Rates of growth of Soviet industrial factor productivity, selected periods, 1961-1982**

Per cent

	Average Annual Rate of Growth[a]								
	1961-65	1966-70	1971-75	1976-80	1978	1979	1980	1981	1982
Industrial production	6.5	6.3	5.9	3.2	3.3	2.1	2.8	2.5	2.2
Inputs									
Man-hours	2.9	3.1	1.5	1.6	1.7	1.3	1.1	0.9	0.7
Capital	11.4	8.8	8.7	7.7	7.0	7.9	7.1	7.8	6.9
Man-hours and capital combined[b]	6.9	5.7	4.9	4.5	4.2	4.4	3.9	4.1	3.6
Factor productivity									
Man-hours	3.5	3.1	4.4	1.6	1.6	0.7	1.7	1.6	1.5
Capital	–4.4	–2.3	–2.6	–4.2	–3.5	–5.4	–4.0	–4.9	–4.3
Man-hours and capital combined	–0.3	0.5	1.0	–1.2	–0.9	–2.2	–1.1	–1.6	–1.3

a) The base year is the year before the stated period.

b) Inputs of man-hours and capital are combined using weights of 52.4 per cent and 47.6 per cent, respectively, in a Cobb-Douglas (linear homogeneous) production function. These weights represent the distribution of labour costs (wages and social insurance deductions) and capital costs (depreciation and a 20 per cent charge on gross fixed capital) in 1970, the base year for all indexes underlying the growth rate calculations.

Source: US Central Intelligence Agency, *Handbook of Economic Statistics 1983* (CPAS 83-10006; Washington, D.C., September 1983), Table 45, p. 66.

study, and it is possible only to note the conclusions of some estimates of Soviet industrial production with constant-elasticity-of-substitution (CES) production functions. Unlike the Cobb-Douglas approach, the CES method does not assume that the elasticity of substitution of capital for labour equals one and the weights of capital and labour are fixed. The CES method thus appears appropriate for a period of Soviet experience when the capital stock grew relative to the labour force. Such CES estimates show that, with capital inputs increasing faster than labour inputs, difficulties in substituting capital for labour caused the diminishing returns to capital responsible for the deceleration in Soviet industrial output growth[5].

The disappointing performance in labour productivity, capital productivity, and combined factor productivity could not be offset by raising the rates of growth of factor inputs. In the case of labour inputs, traditional "reserves" like surplus labour in agriculture and women occupied only in homemaking had been largely exhausted by the early 1950s, and the length of the work-week had been reduced by the mid-1960s. Finally, increments to the total labour force could be expected to decline in the 1980s, compared to the 1960s and 1970s, because of low birth rates in the 1960s, which in turn reflected low birth rates and high death rates during World War II. Nor was it feasible to increase the rate of investment, already absorbing over one-fourth of GNP, in order to raise the rate of growth of the capital stock.

Thus the Soviet regime was compelled to shift from an "extensive" growth strategy, relying chiefly on greater factor inputs, to an "intensive" growth strategy, emphasizing increases in factor productivity. In Soviet publications this shift was often explained in terms of efforts to speed the pace – and capture the fruits – of the "scientific-technical revolution".

Improvements in factor productivity, as measured in production function analyses, can come from many sources. One of them is technological progress in the strict sense – the introduction of new processes and products into production. But the many other possible reasons for changes in output per unit of (combined) factor input include economies of scale; shifts in the allocation of factors from lower- to higher-productivity activities; improvements in organisation, planning, and management at the enterprise, branch, and national levels;

learning from experience; greater effort in response to stronger incentives; weather (affecting the productivity of factors in agriculture); improvements in the quality of labour inputs (conventionally measured in worker-hours) due to educational advance; and higher real costs and lower productivity of labour and capital as less rich, deeper, and more remote natural resources are used.

Although these elements can be identified analytically, it is not possible to estimate empirically the role of each in the overall change in factor productivity and thus to isolate the contribution of technological progress in the strict sense.

For example, a recent ambitious though necessarily rough effort in this direction by Bergson calculated the change in combined factor productivity of labour, capital, and land in Soviet material production (i.e., excluding services) and then estimated the share of the change due to:

1. Labour quality improvements from educational advance;
2. Labour transfers from agriculture to industry;
3. Economies of scale;
4. Weather; and
5. Natural resource exhaustion.

The remainder of the factor productivity change was ascribed to a residual labelled "technological progress proper" (TPP). This includes technological progress in the strict sense but it also encompasses the effects of other elements like learning by experience, changes in effort, and changes in organisation, planning, and management.

Bergson found that the average annual contribution of TPP to the growth rate of combined factor productivity in material production declined sharply from a respectable 2.88 percentage points in the 1950s to 0.98 per cent in the 1960s and only 0.16 per cent in 1971-75. Because they are based on many simplifying assumptions and rough calculations, these estimates can at best offer only a very approximate idea of the rate of TPP in Bergson's sense, of change in it, and of the behaviour of technological progress in the stricter sense[6].

However, such calculations give some support to the view that lagging technological progress has depressed the rates of growth of factor productivity and output in the USSR. This view is also supported by studies of systemic handicaps to technological progress in the USSR, on the one hand, and assessments of the technological lag of Soviet industry behind the West, on the other.

2. Handicaps to Technological Progress in the USSR

It might appear that in principle a socialist centrally planned economy has some advantages over capitalist market economies in the potential for technological progress. For instance, central planning agencies can initiate, organise, co-ordinate, and finance fundamental and applied research, development, and innovation on a national scale, according to offical priorities, and share the results widely without restrictions of commercial secrecy. Also, central planning can enforce a high rate of investment, achieving a rapid expansion of the capital stock that can embody new process and product technologies. Further, this investment programme can be concentrated in branches of the economy with the greatest potential for technological progress – machine building rather than services, for example.

However, the effectiveness of this administrative demand from above for technological progress may be limited by other systemic features that impede technological progress by weakening the response of enterprises and research organisations to such official directives and by discouraging decentralised initiatives for innovation. These features include, for example, problems in organisation, financing, supply, pricing, and performance indicators and incentives[7].

One major organisational handicap is the separation of research and development (R & D) from production. R & D are typically performed in centralised branch research facilities whose task is to develop new products and processes in response to instructions from the branch ministry, which, after accepting the results, assigns them to particular enterprises. The lack of contact between design bureaux and producing enterprises often causes the design bureau to do work unsuited for application in production and even to be indifferent to the fate of its work once accepted by the ministry.

Another problem hampering the introduction of R & D into production is the underfinancing of work at the boundary between research and production, namely, testing, pilot production, and demonstration facilities.

Even when the enterprise is technically capable of introducing an innovation, it may be prevented or discouraged from doing so by the chronic shortage of material inputs in the overstrained Soviet economy. Innovation is likely to exacerbate already serious supply problems for the enterprise because it often requires new and unfamiliar materials and equipment, frequently from newly assigned suppliers which are not acquainted with the enterprise's activities and with which the enterprise lacks the close working relations necessary to secure scheduled deliveries of scarce supplies. Further, innovation may entail unanticipated difficulties in the use of materials and equipment, waste until new technology is mastered, and thus sudden needs for supplies not foreseen when the supply plan for the period was prepared and approved. Supply problems and the consequent risks for plan fulfilment can be reduced if the enterprise produces (more of) its familiar output mix with existing methods, avoiding product or process innovation.

Innovation has also been discouraged by the traditional Soviet policy of setting industrial wholesale prices as the sum of planned average cost of production plus a standard profit markup. An enterprise that could introduce new and cheaper processes has little incentive do so so if cost reductions lead to decreases in output prices and thus tend to depress the value of sales and profits. An enterprise that could produce at the same cost a new model of a machine with greater productivity would get no increase in price, sales, and profits, as the purchaser would capture all the benefits of the innovation in the form of a reduction in the price per unit of use-value. To address this problem, in the late 1960s the pricing scheme for new products was changed to permit "surcharges" to provide an above-average profit markup on average cost. The size of the surcharge thus determines the distribution of gains from innovation between the producer and the user. However, producing enterprises complain that these surcharges are too small and/or remain in effect for too short a period, as the State Committee on Prices strives to exert pressure for cost reduction by decreasing surcharges when profitability is deemed excessive. The industrial wholesale price revision of 1982 applied the surcharge approach to a wider range of products and increased surcharge rates somewhat, but it is too soon to assess the effects on the output of new products.

Despite many changes in the last two decades, the system of performance indicators for Soviet enterprises still emphasizes output – though now marketable output or actually sold output rather than simply gross output – more than other indicators such as cost, profitability in relation to capital, labour productivity, quality, introduction of new technology, savings in use of materials and fuels, etc.

Consider an enterprise director with some choice between seeking (or at least not resisting superior agencies' pressure) to introduce a new product versus continuing to produce a familiar product-mix. If he is completely successful with the new product assignment, he (and other members of the management staff) will receive a new technology bonus in addition to the basic bonuses for output, profitability, etc.

But by innovating the manager incurs significant risks that:

1. The supply problems mentioned above occur;
2. The shakedown time for mastering the output of the new product will be longer than planned;
3. The production cost of the new product will exceed the target;
4. The price surcharge will provide only a modest profit, especially if the planned cost is exceeded;
5. The quantity produced will fall short of plan;
6. The new product will not perform for customers as well as claimed by the R & D organisations that developed it; and
7. Managerial bonuses will be lost.

If the enterprise does not innovate, it is much more likely to fulfil most if not all plan targets and earn managerial bonuses almost as large as those for successful innovation. Assessing the risk-reward relationship, most enterprise managements prefer to avoid innovation.

Further, unlike management in market economies, Soviet managers are not under competitive pressure to introduce new products and processes in order to secure profits, increase market share, and avoid business failure or takeover of the firm. Rather, in a tautly planned economy of widespread and chronic shortages, the Soviet manager is assured that supply agencies will place all the marketable output the enterprise produces.

In turn, the performance of associations and ministries is evaluated by aggregation of the results of their component enterprises. Hence, these higher-level organisations have a similar interest in securing a greater output of the existing product-mix through known methods of production, in preference to the risks of introducing new products and processes. To some extent this attitude can also affect the ministry's view about the importance of branch-sponsored R & D.

These innovation problems help to explain the Soviet lag in the implementation of inventions, as measured by the lead time between the filing of an inventor's certificate and certification of the introduction of the invention. According to a study for the early 1970s[8], lead times are considerably longer in the USSR than in the United States and the Federal Republic of Germany. For instance, in the sample investigated, at the end of two years only one-fourth of the inventions in the USSR were implemented – compared with two-thirds in the United States and the Federal Republic of Germany. In the USSR the mean lead time was 4.26 years in civilian machine building, including computers, instruments, and electrical machinery – key branches for technological progress. The "compounding" effect of successive "generations" of long lead times is one element in the weak technological progress contributing to the sustained decline in Soviet factor productivity discussed above.

In turn, this implementation lag is explained in part by Soviet problems in new plant construction, on which the Soviet authorities have relied – more than replacement of machinery and equipment in existing facilities – as the major channel for embodied technological progress. Plans for output, investment, and commissioning of new capacity are poorly co-ordinated. Construction and installation work progresses unevenly and slowly because of shortages of materials, equipment, and labour, and as a result projects are often completed far behind schedule. Construction organisations still get bonuses mainly for the ruble value of work done, rather than for timely completion. Machine building enterprises normally are not responsible for the delivery and installation of their products, having fulfilled their obligations with the transfer of output to supply agencies. Equipment-installation organisations also perform capital-repair work at existing enterprises, which they favour because bonuses are more easily earned there. Finally, the penalties for slow assimilation of new capacity by the enterprise are modest. The capital charge is usually only 6 per cent, and annual output assignments tend to take assimilation problems into account[9].

The campaigns and "reforms" undertaken to address these problems have failed to overcome them. For example, one ambitious attempt was a broad package of measures adopted in 1968 affecting planning, organisation, incentives, and contracting for research, development, and innovation (RDI)[10].

In RDI planning, the changes included earmarking funds for key scientific and technical problems of economy- and branch-wide significance; long-range technological forecasts; five-year, rather than only annual, RDI plans; and "co-ordination plans" to cover all the work on a single problem from fundamental research through assimilation into production.

The organisational measures sought to break down administrative barriers between RDI activity and production. One method was the combination of various R & D organisations under a single management to develop specific lines of technology. The other was to establish scientific-production associations *(nauchno-proizvodstvennye ob"edineniia* – NPOs) including R & D institutes, design bureaux, experimental plants, and even producing enterprises.

The changes in incentives aimed to relate compensation for RDI to its "economic effect" in production due to cost reduction or output increases from new processes or products. Some of the allocations to economic incentive funds of R & D organisations were linked to the economic effect of their work. Experiments were undertaken to relate individual salary differentials in R & D organisations more closely to performance.

More of RDI was to be financed by contracts (or similar ministerial "work-orders") rather than direct budget allocations to R & D performers, in the belief that contracting reduced the cost and time of R & D and increased its responsiveness to customers' needs.

It is not possible to evaluate statistically the effects of the 1968 RDI measures on Soviet technological progress, but it appears that they have been relatively modest. Many of these measures have been only partially implemented. The specialised Soviet literature still reports many of the shortcomings in RDI that the 1968 reforms were supposed to correct, such as poor co-ordination of RDI plans with production, supply, and investment plans; duplication in the conduct of RDI work; delay or failure in the introduction of innovations (leading sometimes to the obsolescence of new technology before it is applied); and failure to establish a satisfactory system of incentives for the creation and introduction of new technology[11].

Thus, all of the fundamental problems of Soviet RDI remain. Periodically, the Soviet Government asserts its concern about them and repeats its call for efforts to correct them. The most recent example is the August 1983 decree, "On Measures to Accelerate Scientific and Technical Progress in the National Economy"[12].

3. The Soviet Technological Lag

The problems in generating domestic technological progress just discussed help to explain the widely accepted view that in many, though not all, fields Soviet technology is behind that of leading Western countries. However, it is impossible to measure such lags in any single unambiguous way and thus difficult to research summary conclusions[13].

Some international comparisons of technological levels focus on differences in aggregate indicators like the number and use of R & D personnel and the amount and distribution of R & D expenditures. But the available data are seldom comparable across countries, and in any case they describe annual inputs, rather than achieved results. Also of limited use are data on international scientific awards like Nobel Prizes or citations of scientists' work in foreign journals, since they refer to individual achievements, usually in fundamental research rather than in applied research, development, or innovation.

Inventiveness in creating laboratory-scale prototypes or test batches of new products can be approached by examination of patent registrations. But information on the number and content of patent applications or awards sheds little light on the technological level of production, which depends on the successful and wide use of patented (or non-patented) new processes and products.

Thus, the date of the first commercial use of a significant new process or output of an important new product provides a more genuine technological milestone. Still more useful as indicators of technological progress are the speed and extent of diffusion of a new technology, showing the ability to replicate the innovation in other plants and use it on a large scale. However, because diffusion involves replacement investment or increases in capacity, the rate of diffusion depends not only on technological capabilities but also on the rate and composition of investment in an economy.

Once commercial production has been achieved, its technological level may be assessed by comparing technical parameters of, say, a new machine with those of foreign counterparts. Such parameters include precision, speed, weight, energy consumption, reliability in operation, service life, degree of automation, type of process control, and others. However, such comparisons are often ambiguous, as one machine may be superior to another in one parameter but inferior in a different one, so that some weighting of the relative importance of different parameters is necessary to reach an overall evaluation.

Another approach to assessing comparative technological levels is through statistics on international trade in licences and in machinery and equipment – involving, respectively, unembodied and embodied technology. A country's purchases (vs. sales) of licences and machinery and equipment provide useful evidence on relative technological levels, though with some limitations. To some extent imports of licences and of machinery and equipment are substitutes, and the two categories should therefore be considered together. However, as discussed further below in Chapter 3, for reasons of commercial secrecy the information available on licences usually covers their number and general subject, rather than their value and specific content. In contrast, trade figures published by government agencies are much more detailed. The commodity composition of foreign trade can thus help in the assessment of relative technological levels and the actual content of (embodied) technology transfer between countries. Thus, analysis of the commodity composition of Soviet foreign trade with OECD countries shows the much greater share of machinery and equipment in Soviet imports than in Soviet exports (see Chapter 5).

Foreign trade statistics have also been used to calculate average unit values (e.g. in dollars per machine or per kilogramme) in international trade, on the ground that these values will be higher for technologically more sophisticated products. Such calculations indicate lower unit values – and thus an inferred technological inferiority – for Soviet, compared to Western, exports of chemical equipment, machine tools, and synthetic dyes. But the results are sensitive to the choice of unit (the average machine or the average kilogramme), and lower prices may reflect Soviet efforts to break into markets or to compensate for lack of service facilities, rather than (only) technological inferiority.

As the preceding discussion shows, there are various methods of characterising comparative technological levels. Different methods give different results about the extent of the Soviet technological lag for a particular kind of output. In turn, there is no simple way to aggregate (varied) results for individual products into straightforward concise evaluations for the sub-branches and entire branches of industry that have been studied. And no published studies comparing Soviet and Western technological levels are available for some branches, such as clothing and food processing.

Also, international comparisons of technological levels refer to particular points in time and may not be stable over long periods. It is relatively easy to update comparisons based on trade statistics, but detailed analyses of the introduction and utilisation of particular technologies are done infrequently and selectively. As technological change continues in both the USSR and the West, these analyses become out of date, describing the situation ten or twenty years earlier.

Finally, even if one accepts as accurate and contemporary a particular assessment of comparative technological levels, one should have several cautions in mind. As noted above, the Soviet Union may possess a technology through the development stage although it has not been introduced into production, either because introduction was not attempted or because efforts to introduce it have not yet been successful. In the first case, planning agencies may have assigned the activity a low priority (e.g. in consumer goods industry) or have made a specialisation decision to import certain products in order to release domestic resources to other purposes (e.g. military items)[14].

In view of these problems of analysis and interpretation, any comprehensive industry-by-industry survey of evidence about the relative level of Soviet technology would lie far outside the scope of this study. Instead, only a few main findings of some earlier research are summarised here for selected industrial branches in which a technology gap of the USSR behind the West helps to explain the observed pattern of technology transfer (discussed in Chapter 3 below): iron and steel, chemicals, energy, machine tools, motor vehicles, and computers.

Iron and steel[15]. The USSR has been slow to diffuse the oxygen-steel making process, although this may be due to Soviet success in developing giant open-hearth furnaces. The diffusion of continuous casting was much slower in the USSR than the West. The Soviet Union apparently did not undertake direct reduction of iron ore until it signed an agreement with a Federal Republic of Germany consortium in 1975. Although the USSR has vacuum-induction resmelting installations, their utilisation rate is lower than in the West. The volume of Soviet electric furnaces is on average lower than in other major steel-producing countries. Although the electro-slag process is a Soviet innovation widely used abroad, the USSR lags in other refining processes, and in the technology of making tinplate, other types of coated metal, and especially plastic-coated steel. The output of high-quality thin sheet steel (used in light automobiles and domestic appliances) is relatively low, since Soviet rolling mills were developed primarily for the production of sectional steel and thick plate for heavy engineering and construction.

Chemicals[16]. The USSR is self-sufficient in synthetic rubber and simpler chemicals like sulphuric acid. It exports to Eastern Europe relatively simple intermediate products like formaldehyde and urea and imports from the West chiefly more sophisticated products like polyethylene and, especially, artificial fibres. For all the important artificial fibres, production started in the USSR much later than in the West. Beginning in the 1960s, the Soviet Union has imported a great deal of chemical equipment, especially for petrochemicals, plastics, agricultural chemicals (particularly pesticides and complex compound fertilizers), and artificial fibres. The import of complete chemical plants (rather than only licences or equipment) reflects the Soviet Union's weakness in the transition from applied research and pilot-plant production to full-scale mass production.

Energy[17]. The USSR is among the world leaders in high-voltage alternating current transmission of electricity, producing the necessary equipment for stations and networks. It has also been successful in hydroelectric power technology, with generator unit sales to less developed countries and Canada, although some problems occur in generator efficiency and reliability. Soviet nuclear power is also technologically independent of the West, with reactor performance similar to that of Western equipment, but problems have been encountered in the construction of plants. In contrast, Soviet coal-mining technology lags behind that in the West, where, for instance, narrow-web cutters, movable hydraulic roof supports, and advanced drilling equipment were introduced much earlier.

In oil and gas, Soviet technology is clearly behind that of the West, especially the United States. The Soviet Union scored a significant breakthrough in the turbodrill, an indigenous innovation, but further gains have been limited by problems in dealing with greater depth and differences in the kind of rock to be drilled. The USSR has not been able to find satisfactory improvements in new turbodrill designs, electric drilling, better rotary technology, or suitable drill bits. It also lags in the quality of drill pipes, mud technology, and equipment for offshore production, like floating platforms and underwater completion equipment. Soviet exploration technology, including seismic equipment and field computers, can map only about three miles below the surface, compared with ten or more miles for advanced Western equipment. The USSR lags behind the West in the ability to produce large-diameter steel pipe and turbines for compressor stations for gas pipelines. Gas processing technology is deficient in the ability to collect gas at high temperature and pressure, to process gas with sulphur condensate or similar properties, and to provide initial compression for high-volume output[18].

Machine tools[19]. "Conventional" machine tools include traditional types like lathes; drilling and boring, and grinding and milling machines; and transfer lines. In general, Soviet conventional machine tools are similar to Western counterparts in design and operational principles, but inferior in durability, reliability, precision, and flexibility. "Advanced" machine tools include electronics or computers to enhance one or more key aspects, such as flexibility, productivity, or precision. In this category the Soviet Union lags behind the West in the operations that can be performed as well as in the quality of the work, largely because of shortcomings in the electronics available. The USSR introduced numerically-controlled machine tools much later than the West, and when the Soviet machines were produced they were already obsolete by US standards. The gap between US and Soviet machines widened in the 1970s as the United States expanded output of second-generation (transistorised) types and then third-generation types (based on integrated circuits). At the end of the 1970s the Soviet Union was experimenting with direct numerical-control technology, which was available in the United States in the early 1970s but was abandoned in favour of flexible manufacturing systems (FMS) technology. There was no evidence of Soviet development or production of FMS systems.

Among the factors responsible for the Soviet lag in machine tool technology are stress on standardization, the organisation of supply, and the incentive structure. The Soviet machine tool industry emphasizes mass production of highly-standardized, general-purpose machine tools of relatively simple design that are kept in production without major modifications for periods of as much as 15-20 years. In contrast, most US machine tools are specialised – tailored to customer needs – with the attendant stimulus to innovation. In the Soviet Union users have little influence on the quality or design of machine tools allocated to them by supply agencies after their production by largely monopolistic enterprises that are not under competitive market pressure. In turn, the machine building enterprises have quantity and quality problems in the supply of parts and components. As explained above, the incentive structure for machine tool plants and their suppliers favours greater physical output of already "mastered" items rather than quality improvements or new products[20].

Motor vehicles[21]. Innovation in Soviet motor vehicle technology has been very sluggish compared to the West for a number of reasons. The USSR has stressed mass production of vehicles with rigidly standardized designs for very long periods without model changes. This approach was possible because lack of market pressure permitted continued output of vehicles obsolete in many performance characteristics, such as speed, fuel economy, comfort, handling, and safety[22]. However, this approach retarded the retirement of old production equipment, allowed the stagnation of production technology, and eliminated the need for the machine building industry to develop new machine tools for the production of motor vehicles. Hence, when the Soviet leadership decided in the mid-1960s to modernise and expand motor vehicle production, technology transfer from the West was necessary for both the Volga passenger car plant and the Kama truck plant.

Computers. There is a striking technological gap between the USSR and the West in the development and use of computers, by all the standard indicators, including date of first production of successive generations; performance characteristics like access, storage, speed, and printing; software; user services; and size of inventory relative to the size of the economy. To a great extent, Soviet computer technology is copied from Western, especially US, technology, but with a considerable time lag. In technical capabilities and performance, currently produced Soviet general-purpose computers are similar to those marketed in the United States a decade earlier[23].

The most serious current Soviet lags in computer technology concern large-scale scientific computers, large-capacity magnetic disk storage, microchips, ferrite core main memories, and telecommunications hardware[24].

In addition, the development of microprocessors for numerical control of machine tools and for personal computers has been slow. In the case of personal computers, the lag reflects not only technological problems but probably also official preference for centralised computational facilities rather than the accumulation and processing of data by small units or individuals. Soviet production of a first model of a personal computer began on an experimental basis in 1979, series production of a second model started in 1982, and the Agatha model (based on the American Apple II computer) entered series production in 1983. Soviet specialists estimate a need for "not less than" 50 000 personal computers in the USSR – compared with a reported 2 000 000 personal computers in the United States in 1982 and a projected 10 000 000 in the United States in 1985[25].

It is difficult to formulate broad useful generalisations about the extent of the Soviet technological lag behind the West, because of the number of industrial branches to be considered; the variety of possible indicators; and differences in the date, scope, and depth of relevant published studies. Yet one might characterise the situation around 1970 – near the beginning of the Soviet drive to acquire Western technology analysed in this study – approximately as follows[26]:

1. In high-priority science-based industries – like weapons production[27] and the generation and transmission of electricity – the USSR was on balance at the same technological level as leading Western countries.
2. In high-priority industries that are not so research-intensive – such as iron and steel and machine tools – the Soviet Union was relatively successful in scaling-up established technologies (e.g. very large open-hearth furnaces and mass production of standardised machine tools), but slow to introduce and diffuse new technologies (e.g. continuous casting, oxygen-converted and electrical steel, and numerically-controlled machine tools).

3. In somewhat lower-priority science-based industries – like chemicals, computers, and motor vehicles – the Soviet lag was clear and marked.
4. In low-priority traditional branches – like distribution, food-processing, clothing, footwear, and most consumer durables – the gap between the USSR and the West was probably greatest, although comparative technological levels in these branches are not adequately covered in the published literature.

This picture of relative technological lags, together with Soviet economic development priorities, explains the pattern of technology acquisitions by the USSR examined in detail in Chapter 3.

During the period since 1970, technological levels have risen in various parts of Soviet industry, as a result of both domestic technological development and the import of foreign technology. But technological development has, on the whole, proceeded at least as fast, and often faster, in the same industrial branches in Western developed market economies. Thus, the Western and Soviet specialised literature does not indicate any trend toward a narrowing of the general Soviet technological gap discussed above[28]. This continued gap is consistent with, and in large measure explained by, the failure to achieve significant improvements in the Soviet RDI system, discussed in sub-section 2 above.

At the same time, one should recognise that in selected fields the USSR not only matches but possibly leads Western technology. This is shown, for example, by Soviet sales to Western firms of licences for Soviet welding equipment, cooling technology for blast furnace linings, electromagnetic casting of copper, surgical instruments, and other products and processes[29].

B. SOVIET DECISION-MAKING ON THE ACQUISITION OF WESTERN TECHNOLOGY

This section examines some major facets of Soviet decision-making on the acquisition of Western technology. First the chief organisations involved are identified and their roles are described briefly. Then criteria for technology import decisions are examined.

1. Organisations

A large number of Soviet Government agencies are involved in technology acquisition decisions. The most important are the following[30]:

1. The USSR Council of Ministers is the top government policy-making body and resolves disagreements among lower agencies.
2. The USSR State Planning Commission (Gosplan) is responsible for planning imports and their use in the Soviet economy, and also participates in planning R & D and the introduction of new technology of domestic and foreign origin.
3. The State Committee for Science and Technology (SCST) is responsible for Soviet technology policy, including the co-ordination of RDI throughout the economy. This task includes developing strategies for the acquisition of Western technology and integrating it with domestic R & D capabilities. The SCST also is involved in negotiations for particular Western technology acquisitions.
4. The USSR Academy of Sciences monitors specific developments in the West and actively participates in scientific exchanges, one channel of technology acquisition.

5. The Military-Industrial Committee of the USSR Council of Ministers has primary responsibility for co-ordinating armaments production. It and the Ministry of Defence presumably take part in decisions about technology acquisitions with military implications.
6. The State Committee for Material-Technical Supply (Gossnab) must include imports (and exports) in supply plans for branches and enterprises.
7. Each of the forty-odd branch ministries has a main technical administration to formulate and implement technological policies in its sphere.
8. The Ministry of Foreign Trade is involved in many aspects of technology acquisition, through foreign trade enterprises specialised along product lines and the licence trade organisation. The All-Union Chamber of Commerce arranges exhibitions and contacts between foreign firms and Soviet organisations.
9. The Ministry of Finance is concerned with the receipt and expenditure of convertible currency. The Foreign Trade Bank, a unit of the State Bank, handles credits and payments in foreign currencies.

These government agencies have detailed control over the planning, acquisition, and use of foreign technology. However, the Communist Party Secretariat plays a leading role in the formulation of overall policy on technology imports and very likely participates in decisions on specific important acquisitions. Also, republic- and province-level Party officials often take part in particular technology acquisition decisions affecting their territory, especially in the case of the construction of new facilities in which Western equipment will be installed.

2. Decision Criteria

Soviet criteria for decisions on the acquisition of Western technology are in part inferred from the technology lags discussed above; in part explicitly indicated by leaders' policy speeches, material in plan documents and technical literature, and statements by Soviet foreign trade officials; and in part revealed by the actual pattern of Soviet technology imports from the West[31].

Of fundamental importance are the decisions, beginning in the mid-1960s, at the highest political level to acquire Western technology on a significant scale through commercial channels[32]. Then the choice of technologies to be sought from the West depends on the following[33]:

1. Soviet plans to increase output of particular branches and products.
2. The investment – especially new plant construction – necessary to achieve the output increase.
3. The capacity of the Soviet machine building industry to supply the needed quantities of machinery and equipment at the desired technological level.
4. The ability of the USSR to obtain from East European countries what Soviet industry cannot provide.
5. The potential contribution to future Soviet exports for convertible currencies.

These criteria are numerous and clearly will frequently give conflicting results about the desirability of technology imports from the West. Moreover, some criteria are of greater interest to some decision-making agencies than others – for example, output increases to Gosplan vs. convertible-currency costs to the State Bank. But it is unlikely that there is any general scheme of assigning the various criteria ranks or weights. Instead, the criteria represent considerations to be taken into account in interagency negotiations to reach particular technology import decisions[34].

It is noteworthy that formal cost-benefit calculations appear to play little role in Soviet technology import decisions[35]. The main reason is that Soviet internal prices and costs reckoned in them do not usually reflect relative scarcities, and Soviet foreign exchange rates do not correspond to the relative purchasing powers of the ruble and foreign currencies or aim to balance the supply of and demand for foreign currencies. With non-scarcity prices and non-equilibrium exchange rates, Soviet planning agencies cannot accurately measure the domestic costs of investment and production and compare the costs and benefits of domestic vs. foreign technologies. Thus, for instance, formal cost-benefit calculations did not guide decisions on the importation vs. domestic production of pipe for gas pipelines, according to a careful analysis by a leading Western specialist on Soviet energy technology[36].

The application of the criteria mentioned, by the agencies identified in the preceding sub-section, has led to substantial Soviet imports of Western technology for the motor vehicle, chemical, oil and gas, machine building, mining, and timber industries[37]. However, the Soviet technology acquisition effort has been constrained, in different ways, by three factors discussed in the next section.

C. CONSTRAINTS ON SOVIET ACQUISITION OF WESTERN TECHNOLOGY

Among the constraints on the acquisition of Western technology by the USSR are the Soviet ability to assimilate such technology, the Soviet capability to finance technology imports, and Western governments' restrictions on the export of technology to the USSR. Each of these constraints is examined in this section.

1. Soviet Ability to Assimilate Western Technology

The assimilation of technology involves:

1. An absorption phase in which the technology is "mastered" (in Soviet parlance) in the facility in which it is first introduced; and
2. A diffusion phase in which the technology is subsequently applied in other enterprises – showing the ability to replicate the innovation and use it on a wider scale.

The assimilation of Western technology in the Soviet economy encounters difficulties of two kinds. The problems in the introduction of new technology of domestic origin apply also to foreign technology. As discussed in Section A above, these problems include organisation, supply, pricing, performance indicators and incentives, and delays in plant construction. But it is to be expected that the assimilation of Western technology will usually be more difficult than the assimilation of Soviet technology precisely because the Western technology is both foreign and more advanced than Soviet technology. Compared to Soviet technology, Western technology imposes greater, harder-to-satisfy requirements for complementary inputs of Soviet origin, including buildings, machinery and equipment, skilled labour, high-quality materials, and research and development services. When the quantity and/or quality of these complementary Soviet inputs are inadequate, facilities incorporating Western technology are completed behind schedule and, once they are in operation, production does not meet output and quality targets[38].

Assimilation problems experienced in the importation of Western technology for particular branches of Soviet industry are discussed below in Chapter 4 in connection with the assessment of the impact of technology transfer from the West on the Soviet economy. The recognition of these problems has influenced Soviet decision-makers' views on the advisable nature, amount, and rate of technology imports from the West.

2. Soviet Financing Capability

The Soviet Union's ability to finance imports of Western technology depends primarily on three related factors. One is the excess of convertible currency earnings from exports over outlays for other high-priority imports like grain. Another is the availability and terms of Western credits to cover deficits in the current account. The third is the extent of Soviet success in making countertrade arrangements for future exports to service the convertible currency debt. Each factor will be discussed in turn.

a) *Soviet Convertible Currency Balance of Payments*

The USSR does not publish statistics on its balance of payments either with the rest of the world as a whole or with particular regions or currency areas[39]. Hence, analysis of the structure and behaviour of the Soviet convertible currency balance of payments must rest on published Western estimates, which are necessarily incomplete, are not always mutually consistent, and are subject to revision. Table 6 presents recent Western estimates for the Soviet convertible currency balance of payments in 1970, 1975, and 1978-1982[40]. The figures cover Soviet transactions in convertible currencies not only with developed market economies but also with non-Communist developing countries and (on a small scale) with other Communist countries. Also, it should be noted that in the format of Table 6 Soviet arms sales are included in "other" current account transactions, rather than in merchandise exports.

As Table 6 shows, in convertible currency transactions Soviet merchandise imports exceed merchandise exports, causing a persistent deficit in merchandise trade. In addition, net interest payments are negative. Offsetting the trade and interest income deficits are arms sales to developing market economies, gold sales, and net borrowing abroad (in most years). Preliminary data for Soviet convertible currency transactions in 1983 indicate a similar picture. An increase in the merchandise trade deficit (to about $2.8 billion) was offset by larger arms sales and net borrowing abroad (in contrast to net repayments in 1982).

Table 6. **Soviet convertible currency balance of payments, 1970 and 1975-1982**

Millions of US dollars

	1970	1975	1978	1979	1980	1981	1982
Current account balance	260	–4 607	425	2 177	1 904	–100	4 206
Balance on merchandise trade	–560	–6 297	–3 690	–2 018	–2 486	–4 000	–1 294
Exports, fob	2 424	8 280	13 336	19 417	23 584	23 778	26 152
Imports, fob	2 984	14 577	17 026	21 435	26 070	27 778	27 446
Net interest	–80	–570	–880	–800	–710	–1 300	–1 500
Other transactions[a]	900	2 260	4 995	4 995	5 100	5 200	7 000
Capital account balance	265	6 520	1 735	340	1 630	5 840	–1 240
Net borrowing from abroad[b]	290	5 400	765	1 675	–185	3 000	–765
Net change in assets[c]	–25	395	–1 550	–2 825	235	140	–1 575
Gold sales	0	725	2 520	1 490	1 580	2 700	1 100
Net errors and omissions[d]	–525	–1 913	–2 160	–2 517	–3 534	–5 740	–2 966

a) Including receipts from arms sales, official transfers, and net receipts from tourism and transportation.
b) Soviet drawings on Western credits less repayments on them.
c) Net change in Soviet assets held in Western commercial banks.
d) Including Soviet convertible currency aid to, and trade with, other CMEA countries; and trade credits to finance Soviet exports to non-Communist convertible currency trade partners.
Source: US Central Intelligence Agency, *Handbook of Economic Statistics, 1983* (CPAS 83-10006 ; Washington, D.C., September 1983), Table 49, p. 70.

Within total Soviet merchandise imports for convertible currencies, purchases of Western machinery and equipment grew rapidly after 1970, increasing sixfold in 1971-1978 to about $6 billion. The rise in these imports was particularly fast in 1975, when deliveries almost doubled over the previous year, reaching $4.6 billion. Shipments in 1976-1978 averaged about $5.4 billion per year. This dramatic growth ended in 1978. Exports of capital goods from OECD countries to the USSR in 1979 were about the same as in 1978 in value terms and, when price increases are taken into account, less in real terms. In both nominal and real terms, such exports to the USSR declined in 1980 and 1981, but rose in 1982 (see Chapter 3 and Annex Tables A-6 and A-7).

However, Soviet decisions to curtail, and subsequently to increase, technology imports from the West were made somewhat earlier than the trade figures indicate, since deliveries of machinery and equipment usually lag behind the corresponding orders by one-two years. Thus, for instance, the large imports of machinery and equipment in 1976-1978 resulted from orders placed in 1974-1976. Such orders declined from about $6 billion in 1976 to about $3.8 billion in 1977. They have averaged $2-3 billion annually since 1979[41].

Changes in Soviet technology imports from the West are explained by a complex combination of factors. For example, Soviet ability to pay for Western machinery, equipment, and intermediate products like large-diameter pipe is enhanced by several kinds of developments. They include improvements in Soviet terms of trade, due largely to movements in world oil prices; higher prices on world gold markets; and greater opportunities for arms sales to Third World countries.

On the other hand, Soviet purchases of Western capital goods tend to be curtailed by other factors. Among them, for instance, are additional grain imports in years of bad harvests in the USSR; principal and interest payments to service a growing Soviet convertible currency debt; and problems encountered in assimilating Western technology[42].

In some years, for example, 1977 and 1981, the USSR covered part of its convertible currency merchandise trade deficit by drawing on its reserves of gold and convertible foreign exchange (see Table 7). According to the data in Table 8, Soviet gold production has grown steadily since 1970 (except in 1975). But during 1976-1978 the USSR reduced its gold reserves by 20 per cent, from 61.08 million troy ounces at the end of 1975 to 49.08 million troy

Table 7. **Soviet international reserves of gold and convertible foreign exchange, end of year, 1970 and 1975-1981**

Millions of US dollars

Year	Amount
1970	2 800
1975	5 628
1976	7 083
1977	6 815
1978	8 213
1979	11 114
1980	11 297
1981	11 100

Source: US Central Intelligence Agency, *Handbook of Economic Statistics 1981* (NF HES 81-001; Washington, D.C., November 1981), Table 18, p. 36 and *Handbook of Economic Statistics, 1983* (CPAS 83-10006; Washington, D.C., September 1983), Table 18, p. 41.

Table 8. **Soviet gold production and reserves, 1970-1982**

Millions of Troy Ounces

Year	Production During Year	Reserves End of Year
1970	7.00	52.43
1971	7.20	57.77
1972	7.81	59.22
1973	8.03	56.16
1974	8.42	59.02
1975	8.29	61.08
1976	8.87	57.77
1977	9.19	54.72
1978	9.54	49.08
1979	9.87	50.83
1980	10.20	58.22
1981	10.50	60.72
1982	10.60	68.10

Source: US Central Intelligence Agency, *Handbook of Economic Statistics 1983* (CPAS 83-10006; Washington, D.C., September 1983), Table 51, p. 71.

ounces at the end of 1978. Gold reserves were then rebuilt, especially in 1980 and 1982. At the end of 1982 they amounted to 68.10 million troy ounces.

b) *Western Credits*

Western credits to the USSR were a major source for covering Soviet trade and current account deficits.

However, despite the importance, to both sides, of the Soviet debt to the West, the pertinent statistical information available is deficient. The USSR publishes very little relevant data, and the statistical reports of Western governments and international organisations are incomplete[43]. As a result, even the most elaborate and careful Western estimates of the Soviet debt are imprecise and vary considerably in:

1. What kinds of official and private Western credit are included;
2. The distinctions made among commitments, actual drawings, and obligations outstanding after repayments;
3. The use of gross or net liabilities of the USSR to Western banks;
4. Emphasis on stock figures (like liabilities at a given date) versus flow figures (such as loan drawings over a particular period);
5. Attention to the maturity distribution of outstanding obligations;
6. Treatment of CMEA banks' indebtedness (sometimes stated separately, sometimes included in the figures for the USSR, and sometimes ignored);
7. Possible double counting, for example of bank holdings of supplier credit paper and government-guaranteed debt; and
8. Treatment of short-term securities of Western governments held by the USSR outside Western banks.

Thus, Western estimates like those in Table 9 are based on incomplete data, involve simplifying assumptions, are not always internally consistent, have a considerable margin of error, are subject to revision, and may differ significantly from other Western estimates (even those made by the same organisation earlier or by other methods)[44].

Table 9 presents the most comprehensive estimates publicly available for Soviet convertible currency debt and debt service to the West in 1971 and 1975-1982. Soviet gross

Table 9. **Soviet convertible currency debt and debt service to the West, 1971 and 1975-1982**

	1971	1975	1976	1977	1978	1979	1980	1981	1982[a]
Debt and offsetting assets at end of year (billions of US dollars)									
Gross debt	1.8	10.5	14.7	15.6	16.4	18.1	17.8	20.9	20.1
Commercial debt	0.4	6.9	9.7	9.8	9.5	10.5	10.0	13.0	11.5
Medium- and long-term	*b*	*b*	*b*	*b*	6.4	7.2	6.1	7.5	7.5
Short term	*b*	*b*	*b*	*b*	3.1	3.3	3.9	5.5	4.0
Government-backed debt	1.4	3.6	5.0	5.8	6.9	7.6	7.8	7.9	8.6
Assets with Western banks	1.2	3.1	4.7	4.4	6.0	8.8	8.6	8.4	10.0
Net debt	0.6	7.4	10.0	11.2	10.4	9.3	9.2	12.5	10.1
Debt service during year (billions of US dollars)	0.3	1.8	2.4	3.1	3.6	4.2	4.7	5.4	5.6
Debt-service ratio (per cent)[c]	10	15	16	17	17	16	15	17	16

a) Preliminary estimate.
b) Not available.
c) Debt service divided by the sum of receipts from merchandise exports, arms sales, gold sales, interest, net invisibles, and transfers.
Source: US Central Intelligence Agency, *Handbook of Economic Statistics, 1983* (CPAS 83-1006; Washington, D.C., September 1983), Table 50, p. 71.

convertible currency debt rose almost fivefold from 1971 to 1975, and then increased another 40 per cent in 1976. Thereafter, Soviet gross debt has grown in some years (1977-1979 and 1981) and decreased in other years (1980 and 1982).

Although, according to Table 9, credits backed by Western governments have accounted for less than half of Soviet gross convertible currency debt, such credits are an important source of financing Soviet imports of machinery and equipment, especially for large projects. From the Soviet viewpoint, these officially-supported credits have several advantages over commercial debt. In addition to bearing the political endorsement of Western governments, these loans have long maturities at fixed rates of interest well below fluctuating market rates on commercial debt[45].

However, to obtain the Soviet net debt position, one must deduct from Soviet gross debt the partially offsetting Soviet assets in Western banks. The magnitude of these Soviet assets is significant. For instance, according to the data in Table 9, at the end of 1982 these Soviet assets, amounting to $10.0 billion, were equal to about half of the Soviet gross debt of $20.1 billion.

Service on the Soviet convertible currency debt – the sum of principal repayment and interest – has grown steadily, reaching $5.6 billion in 1982. The "burden" of debt service is commonly reckoned as the ratio of debt service to some combination of earnings potentially available to make debt service payments. The last row in Table 9 shows Soviet convertible currency debt service as a percentage of the sum of Soviet convertible currency receipts from merchandise exports, arms sales, gold sales, interest payments, net invisible transactions, and transfers. The USSR has maintained this ratio at 15-17 per cent since 1975. Debt service ratios of about 25 per cent in relation to merchandise exports and about 20 per cent in relation to total current receipts are usually considered reasonable and prudent. The Soviet Union's willingness to curb imports, on the one hand, and its ability to earn convertible currency from exports of fuels, raw materials, arms, and gold, on the other, have led to generally favourable Western evaluations of the "creditworthiness" of the USSR[46].

c) *Countertrade*

One aspect of Soviet conservatism in financing imports is the emphasis on countertrade in the form of long-term compensation agreements. Through these arrangements, the USSR links the amount and the timing of the repayment of credits for imports of Western machinery and equipment to a contracted flow of convertible currency exports from the projects incorporating the imports.

Various types of East-West countertrade can be identified, according to such dimensions as:

1. The number and types of contracts;
2. The nature, origin, and amount of products supplied by the Eastern country;
3. The nature and extent of the relationship between the Western export and the Eastern export;
4. The number of parties involved besides the buyer and the seller;
5. The "counterdelivery ratio" expressing the value of Western purchases as a percentage of Eastern purchases;
6. The extent to which the Eastern export is used by the Western exporter;
7. The amount of time to complete the transactions involved; and
8. Financing procedures.

By these criteria, three main categories of countertrade are usually distinguished[47].

Barter agreements involve a single contract for a direct exchange between two parties of goods having offsetting values, without the flow of money, usually over a relatively short period of time (up to two years). Such contracts are very rare in East-West trade.

Counterpurchase agreements cover deals, over a 1-5 year period, for machinery, semimanufactures, raw materials, or consumer goods in which the Eastern exports have no intrinsic link to the Western exports. Two legally separate contracts provide for cash payment by each side for products and services received. However, financing from Western banks may assist Eastern purchases, and the Western firm may not use the Eastern goods itself but instead may transfer them to a trading house for resale.

Compensation agreements involve longer-term (5-10 or sometimes more years) deals of relatively large value (hundreds of millions, or sometimes billions, of dollars). They provide for Eastern imports of machinery and equipment and other forms of technology from the West largely on credit, with subsequent repayment commonly in "resultant" products produced with the Western-supplied plant, equipment, or technology. The Western "buyback" of "reciprocal deliveries" usually begins at least several years after the Western exports, and the cumulative value of Western purchases may equal or even exceed the value of Western exports (i.e., for a counterdelivery ratio of 100 per cent or more). These "product payback" deliveries are usually valued (annually) at the then prevailing world market prices. Separate contracts cover the Eastern purchase of the Western equipment, credit for it, and the Western purchase of the Eastern product. The Eastern obligation to repay Western credits is not conditional on the progress of counterdeliveries. Although these shipments earn convertible currencies which can be allocated to debt service, foreign exchange from any other source (exports of the same products from other faciltities, other exports, or new credits) can be used for this purpose.

Such compensation agreements have several very desirable features from the Soviet viewpoint[48].

First, Western technology is imported without any outlay of convertible currency, since it is financed by a capital transfer to the USSR.

Also, the loan is "self-liquidating", as repayment, deferred until the project is in operation, is made through guaranteed buyback of resultant product. This assurance is very appealing to Soviet Government agencies concerned about the difficulty of planning exports to the West under conditions of Western business cycles[49]. Further, compensation agreements are often intended to generate exports for larger amounts than needed for debt service alone, and to establish a Soviet position in markets to be served long after the compensation agreement expires.

Further, because Western partners expect Soviet counterdeliveries – at least sufficient to service the debt – Western firms are motivated to work for the successful transfer of Western technology in the form of blueprints, know-how, machinery and equipment, and materials (like large-diameter pipe), including sometimes the updating of technology.

Finally, unlike less developed market economies, the USSR gets the benefits of technology transfer from developed market economies without permitting foreign direct investment in the form of equity participation in the Soviet project.

Soviet compensation agreements have a number of characteristics[50].

1. In size, they are usually for large projects, costing hundreds of millions or sometimes billions of rubles (or dollars).
2. They have a long time span, frequently 10-20 years.
3. By industrial branch, the agreements are concentrated in natural gas (especially Western pipe and pipeline equipment in return for Soviet gas), chemicals, forestry

products, and metallurgy. Table A-1 in the Annex gives a very detailed listing, by branch of industry, for a large number of examples of Soviet-Western compensation agreements, including the Western country and the Western firm; the year of signature; the type and value of Soviet imports; the type, initial date, and value of Soviet exports; and other information.

4. By purpose, Soviet compensation agreements usually entail the construction of new facilities or the development of mineral deposits or other natural resources, rather than the expansion, modernisation, or reconstruction of existing facilities. Although some compensation agreements are for existing facilities, this occurs in only a few branches (for example, the timber industry), the number of enterprises involved is small, their size is small, and the value of the Western equipment deliveries in each case is small.
5. In import composition, all Soviet compensation agreements involve the acquisition of machinery and equipment, often with licences and know-how. In addition, materials for construction and assembly may be imported. The machinery and equipment and materials may be used for infrastructure activities necessary for the project, as well as for direct production facilities. Industrial consumer goods sometimes account for 10-20 per cent of the imports on credit. They may be earmarked for workers on the compensation project, or sold to others through the retail trade network. Receipts from sales of imported consumer goods are viewed as a way of covering some of the ruble costs of construction and assembly work on the compensation project.
6. In connection with credit, Soviet negotiators bargain hard with potential suppliers to get the most favourable terms. The amount of credit depends on the cost and nature of the project. Usually it is at least 40 per cent of the estimated cost, but sometimes it equals the total cost (for instance, in some turnkey projects), and it may exceed the total cost if the goods to be imported are expected to have an especially strong effect on production. Repayment through Soviet exports commonly begins in about 5 years, and may extend for 5, 10, or more years.
7. So far as the purpose of compensation projects is concerned, ordinarily only about 20-30 per cent of the output is exported, under the compensation agreement or otherwise, and the rest used in the Soviet economy. However, the export share is larger for some projects.
8. The minimum counterdelivery ratios required in Soviet compensation agreements rose from 20-30 per cent in the mid-1970s to 30-60 per cent by 1980. For some Soviet exports to the West, such as natural gas, the actual counterdelivery ratio exceeds 100 per cent; for some chemical products it is about 100 per cent; but for Soviet manufactured goods the ratio is much lower[51].
9. Because international trade statistics do not identify separately imports and exports linked to compensation, it is not possible to calculate the share of compensation-based trade in total Soviet commodity trade with the West. Various Soviet published statements in the mid-1970s mentioned figures from 15 to 30 per cent for the share of compensation trade in total Soviet trade turnover (exports plus imports) with the West, but it is not clear to what time period they referred and whether they were plan targets or actual results. More precise is the estimate of trading houses in the Federal Republic of Germany that countertrade was responsible for about 15 per cent of total Soviet trade turnover with the West in 1978[52]. A 1983 Soviet source stated that "in recent years" compensation trade constituted about 13 per cent of Soviet trade turnover with the West[53].

10. In contrast to East European countries, the USSR has made relatively little use of counterpurchase agreements involving exports of Soviet manufactured producer or consumer goods to the West. This reflects the deficiencies of Soviet manufactures – in comparison with those of some East European countries as well as those of the West – in assortment, performance characteristics, style, availability of spare parts and service, and other aspects.

The preceding discussion has examined the Soviet use of countertrade, particularly compensation agreements, as a means to finance Soviet imports of Western technology. The role of countertrade in Soviet exports to the West is considered further in Chapter 5.

3. Western Governments' Export Restrictions

The Soviet Union's acquisition of certain kinds of Western technology has been constrained by unilateral and multilateral export restrictions of Western governments. A detailed treatment of these restrictions is outside the scope of this study, which does not focus on policies of Western governments toward technology transfer to the USSR. However, a brief discussion of the objectives, methods, and effects of these restrictions is appropriate[54].

Particular Western governments may have one or more of the following objectives in restricting exports to the USSR:

1. Denying the USSR technology, for example for guidance and control, with direct military significance;
2. Curbing Soviet acquisition of Western dual (civilian and military) use technology, for instance in computers, of potential military application;
3. Retarding the development of branches, like oil and gas, that play a key role in Soviet economic development and exports; and
4. Protesting specific Soviet foreign policy actions.

The Co-ordinating Committee for Multilateral Export Controls (CoCom) co-ordinates national export controls of the North Atlantic Treaty Organisation countries (except Iceland and Spain) plus Japan. CoCom maintains three lists of products and technologies whose export to the USSR and other Communist countries is embargoed, controlled, or monitored, through export licencing by CoCom member governments. The Munitions List includes clearly military items. The Atomic Energy List comprises sources of fissionable materials, nuclear reactors, and their components. The Industrial List contains other items with civilian uses that also have military or strategic implications. This list includes thousands of items grouped in the following categories:

1. Metalworking machinery;
2. Chemical and petroleum equipment;
3. Electrical and power-generating equipment;
4. General industrial equipment;
5. Transportation equipment;
6. Electronic and precision instruments;
7. Metals, minerals, and their manufactures;
8. Chemicals and metalloids;
9. Petroleum products; and
10. Rubber and rubber products.

However, there are some disagreements among the member governments of CoCom about which specific products or technologies should be added to, or removed from, these lists[55].

Relatively little is published about the many decisions underlying the inclusion or removal of particular items on the lists, the granting of exceptions to them, or the details of rejected licence applications. However, with reference to the focus of this study, one may note that the lists and associated licencing procedures are used to restrict Western exports of selected items for which there is a definite Soviet technology lag and the West wishes for strategic reasons to prevent the USSR from narrowing this gap through the import of Western technology. In this connection, Western restrictions sometimes aim to curb the export of production know-how rather (or more) than sales of final products.

It is clear that the export restrictions adopted by CoCom or by individual Western countries have curtailed Soviet acquisition of Western technology for important branches of Soviet industry such as electronics, electrical machinery, metalworking machinery, oil and gas equipment, and chemical equipment. But it is not possible to reach a quantitative assessment of the impact of these controls. First, only very incomplete information is publicly available on the applications for export licences to the USSR that were denied. Second, as reported in the press from time to time, the USSR has sometimes been successful in evading these controls and obtaining embargoed items, for example through roundabout transshipments. Finally, one does not know to what extent the actual pattern of Soviet technology imports from the West, discussed in the next chapter, involved more or less satisfactory (from the Soviet viewpoint) substitutes for technology denied by Western governments.

NOTES AND REFERENCES

1. Recent comprehensive volumes include John R. Thomas and Urusula Kruse-Vaucienne (eds.), *Soviet Science and Technology: Domestic and Foreign Perspectives* (Washington, D.C.: George Washington University, 1977); Joseph S. Berliner, *The Innovation Decision in Soviet Industry* (Cambridge, Mass.: MIT Press, 1976); Ronald Amann, Julian Cooper, and R.W. Davies (eds). *The Technological Level of Soviet Industry* (New Haven, Conn.: Yale University Press, 1977); and Ronald Amann and Julian Cooper (eds.), *Industrial Innovation in the Soviet Union* (New Haven, Conn.: Yale University Press, 1982).
2. For detailed treatments, see, for example, Stanley H. Cohn, *Economic Development in the Soviet Union* (Lexington, Mass.: D.C. Heath and Co., 1969), and Herbert Block, "Soviet Economic Performance in a Global Context", in US Congress, Joint Economic Committee, *Soviet Economy in a Time of Change,* 96th Congress, 1st Session (Washington, D.C.: US Government Printing Office, 1979,), Vol. 1, pp. 110-140.
3. On these and other problems of official Soviet economic statistics, see Vladimir G. Treml and John P. Hardt (eds.), *Soviet Economic Statistics* (Durham, N.C.: Duke University Press, 1972).
4. See, for example, the selective summary in Ronald Amann, "Some Approaches to the Comparative Assessment of Soviet Technology: Its Level and Rate of Development", in Amann, Cooper, and Davies, *The Technological Level,* pp. 16-19.
5. See Hanson, *Trade,* pp. 31-37, for a more technical survey and appraisal of alternative production function explanations of the retardation in Soviet economic growth.
6. For a fuller explanation of these calculations and their limitations, see Abram Bergson, "Soviet Technological Progress: Trends and Prospects", in Abram Bergson and Herbert S. Levine (eds.), *The Soviet Economy Toward the Year 2000* (London: Allen & Unwin, 1983), pp. 34-78.

7. These are analysed at length in Berliner, *The Innovation Decision,* on which the following discussion is based.
8. John Martens and John Young, "Soviet Implementation of Domestic Inventions: First Results", in US Congress, Joint Economic Committee, *Soviet Economy in a Time of Change,* Vol. 1, pp. 472-509. There is no indication in the specialised Soviet literature of any significant change in lead times in the period since the mid-1970s.
9. Hanson, *Trade,* pp. 66-72.
10. Louvan E. Nolting, *The 1968 Reform of Scientific Research, Development, and Innovation in the USSR* (Foreign Economic Report No. 11; Washington, D.C.: US Department of Commerce, Bureau of Economic Analysis, 1976).
11. The implementation of various measures to improve RDI is evaluated in Julian Cooper, "Innovation for Innovation in Soviet Industry", in Amann and Cooper, *Industrial Innovation,* pp. 453-513; Paul Cocks, "Organising for Technological Innovation in the 1980s", in Gregory Guroff and Fred Carstensen (eds.), *Entrepreneurship in Imperial Russia and the Soviet Union* (Princeton: Princeton University Press, 1983), pp. 318-335; and Morris Bornstein, "Pricing Research and Development Services in the USSR", *Research Policy,* Vol. 13 (1984), No. 2, pp. 85-100.
12. *Pravda, 28th August 1983, p. 1.* The decree notes that

    ... ministries, departments, the USSR Academy of Sciences, and the State Committee for Science and Technology are not displaying the proper persistence in implementing a unified scientific and technical policy ...

    The State Committee for Science and Technology is not fully exercising the rights it has been given to co-ordinate and supervise the organisation of scientific research and the introduction of its results into production. The State Standards Committee is not showing the proper exactingness toward ministries and departments for the quality of development work and the strict observance of standards, and it is not exercising adequate supervision over the quality of output.

    The existing system for evaluating the results of the economic activity of enterprises and organisations is not exerting a sufficiently effective influence on accelerating the creation of new equipment, materials, and production processes ...

    The decree instructs relevant organisations to introduce various relatively modest changes in administrative organisation, planning, performance indicators, pricing, and incentives affecting RDI.
13. The following brief discussion draws on Amann, "Some Approaches", in Amann, Cooper, and Davies, *The Technological Level,* pp. 23-33; Zaleski and Wienert, *Technology Transfer,* pp. 215-218; Josef C. Brada, "Technology Transfer between the United States and the Communist Countries", in Robert Hawkins and A.J. Prasad (eds.), *Technology Transfer and Economic Development* (Greenwich, Conn.: JAI Press, 1981), pp. 233-39; and John Kiser, "Soviet Technology: The Perception Gap", *Mechanical Engineering,* Vol. 101, No. 4 (April 1979), pp. 22-29.
14. These points are stressed by Kiser, "Soviet Technology", pp. 26-27, and Antony C. Sutton, *Western Technology and Soviet Economic Development, 1945 to 1965* (Stanford, Calif.: Hoover Institution Press, 1973), pp. 372-375.
15. See Julian Cooper, "Iron and Steel", in Amann, Cooper, and Davies, *The Technological Level,* pp. 83-120.
16. See Ronald Amann, "The Chemical Industry: Its Level of Modernity and Technological Sophistication", in Amann, Cooper, and Davies, *The Technological Level,* pp. 227-327, and Ronald Amann, "The Soviet Chemicalisation Drive and the Problem of Innovation", in Amann and Cooper, *Industrial Innovation,* pp. 127-211.
17. See Robert Campbell, "Technological Levels in the Soviet Energy Sector", in *East-West Technological Co-operation* (Brussels: NATO, 1977), pp. 241-263.

18. US Central Intelligence Agency, *USSR: Development of the Gas Industry* (ER 78-10393; Washington, D.C., July 1978), pp. 23-24.

19. See James Grant, "Soviet Machine Tools: Lagging Technology and Rising Imports", in US Congress, Joint Economic Committee, *Soviet Economy in a Time of Change,* Vol. 1, pp. 561-572; M.J. Berry and Julian Cooper, "Machine Tools", in Amann, Cooper, and Davies, *The Technological Level,* pp. 121-198; and M.J. Berry, "Towards an Understanding of R & D and Innovation in a Planned Economy: The Experience of the Machine Tool Industry", in Amann and Cooper, *Industrial Innovation,* pp. 39-100.

20. A recent symposium of Soviet specialists on the machine tool industry evaluated the USSR's technological lag and need to import Western technology as follows:

    > ... Soviet machine tools are deficient in many of the characteristics by which their technical level is defined (the most important such characteristic being the degree of automation) ... It's unlikely that any radical progress will be made in this area during the current five-year plan [for 1981-85].
    >
    > This means that billions of rubles must be spent to import many types of machine tools, forge and press machines, and production lines, especially those that are needed for the priority branches of the economy, as well as components and assemblies.
    >
    > Specialists believe that Soviet-made programmed-control devices are technically inferior. They tend to have a limited range of functions and to be large in size, high in cost, and not very dependable. As a result, nearly half of our numerically-programmed multifunctional machine tools, such as the so-called "machining centres", are outfitted with imported components. Obviously, our electrical equipment and electronics industries need to do a great deal more to meet the needs of the machine tool industry.
    >
    > Both Soviet machine tool users and representatives of our export organisations say that Soviet machine tools are considerably inferior to the best world models.

    S.A. Kheinman, "Proizvodstvennyi apparat mashinostroeniia i stankostroenie" [The machinery industry's production apparatus and the machine tool industry], *Ekonomika i organizatsiia promyshlennogo proizvodstva,* 1982, No. 1, pp. 36-40. The views of individual specialists from various ministries are reported in "Vokrug stanka" [Around the machine tool], *Ibid.*, pp. 47-88. An English-language abstract of the two articles appears in *The Current Digest of the Soviet Press,* Vol. 34, No. 18 (2nd June 1982), pp. 5-9.

21. See US Central Intelligence Agency, *USSR: Role of Foreign Technology in the Development of the Motor Vehicle Industry* (ER79-10571; Washington, D.C., October 1979), and Holliday, *Technology Transfer,* ch. 6.

22. For example, Soviet production of front-wheel drive passenger cars is scheduled to begin only in 1985, according to a member of the Design and Experimental Operations Board of the USSR Ministry of the Automotive Industry. Anatoly Titov, "An Insider's Assessment of the Soviet Auto Industry", *The Ann Arbor News* [Ann Arbor, Michigan, USA], 23rd October 1983, p. D10.

23. See Kenneth Tasky, "Soviet Technology Gap and Dependence on the West: The Case of Computers", in US Congress, Joint Economic Committee, *Soviet Economy in a Time of Change,* Vol. 1, pp. 510-523; Martin Cave, "Computer Technology", in Amann, Cooper, and Davies, *The Technological Level,* pp. 377-406; and Seymour E. Goodman. "Soviet Computing and Technology Transfer: An Overview", *World Politics,* Vol. 31, No. 4 (July 1979), pp. 539-570.

24. Seymour E. Goodman, "The Impact of US Export Controls on the Soviet Computer Industry" (unpublished conference paper, March 1983), and Seymour E. Goodman, "Socialist Technological Integration: The Case of the East European Computer Industries", *The Information Society,* Vol. 3 (1984), No. 1.

25. G. Kochetkov, "Dialog 'Chelovek-EVM': Personal'nye kompiutery" [The man-electronic computing machine dialog: Personal computers], *Voprosy ekonomiki,* 1983, No. 4, pp. 152-153. See also "A Russian Computer with an Apple at Its Core", *Business Week,* 21st March 1983, p. 45.

26. Hanson, *Trade,* pp. 41-42.

27. On weapons technology, which is outside the scope of this study, see, for example, David Holloway, "Military Technology", in Amann, Cooper, and Davies, *The Technological Level,* pp. 407-489, and David Holloway, "Innovation in the Defence Sector" and "Innovation in the Defence Sector: Battle Tanks and ICBMs", in Amann and Cooper, *Industrial Innovation,* pp. 276-414.

28. For instance, the USSR is still attempting to master the production of electrostatic copying machines and roll microfilming equipment. The USSR makes no microfiche equipment at all, although microfiche, used for 30 years in the West, is considered the progressive form of microfilm-based information. R. Ivanov, "Vialyi start: O problemakh reprografii – 'skoropisi veka'" [A poor start: On the problems of reprography – the "shorthand of the century"], *Pravda,* 13th March 1984, p. 3.

29. John W. Kiser III, "Tapping Eastern Bloc Technology, *"Harvard Business Review",* Vol. 60, No. 2 (March-April 1982), pp. 85-93.

30. US Congress, Office of Technology Assessment, *Technology and East-West Trade* (Washington, D.C.: US Government Printing Office, 1979), pp. 211-215, gives a brief overview. For a more detailed account, see H. Stephen Gardner, *Soviet Foreign Trade: The Decision Process* (Boston: Kluwer-Nijhoff Publishing, 1983), chs. 2-4.

31. For a historical treatment of the evolution of the views of Soviet officials, scientists, and economists about the importance of technological progress and the desirability of importing foreign technology, see Bruce Parrott, *Politics and Technology in the Soviet Union* (Cambridge, Mass.: MIT Press, 1983).

32. For a review of such decisions, see Zaleski and Wienert, *Technology Transfer,* pp. 154-157, and Hanson, *Trade,* p. 85.

33. See US Congress, Office of Technology Assessment, *Technology,* pp. 217-218, and Chistopher E. Stowell, *Soviet Industrial Import Priorities, With Marketing Considerations for Exporting to the USSR* (New York: Praeger Publishers, 1975), pp. 70-75.

34. For a case study of the interaction of Soviet Government agencies involved in decision-making on natural gas production and transportation, including the acquisition of Western technology, see Thane Gustafson, *The Soviet Gas Campaign: Politics and Policy in Soviet Decision-making* (R-3036-AF; Santa Monica, Calif.: Rand Corporation, 1983).

35. Gardner, *Soviet Foreign Trade,* ch. 5.

36. Robert W. Campbell, *Soviet Energy Technologies: Planning, Policy, Research, and Development* (Bloomington, Ind.: Indiana University Press, 1980), pp. 225-230.

37. Georges Sokoloff, "Politique soviétique d'importation de biens d'équipement: motivations générales et raisons spécifiques de l'appel aux pays occidentaux", *Revue d'Études Comparatives Est-Ouest,* Vol. X, No. 4 (décembre 1979), pp. 117-132.

38. Dennis J. Barclay, "USSR: The Role of Compensation Agreements in Trade with the West", in US Congress, Joint Economic Committee, *Soviet Economy in a Time of Change,* Vol. 2, p. 471.

39. Soviet trade by geographical areas is discussed in Chapter 5.

40. For different estimates, which include other years in the 1970s, see Paul G. Ericson and Ronald S. Miller, "Soviet Foreign Economic Behaviour: A Balance of Payments Perspective", in US Congress, Joint Economic Committee, *Soviet Economy in a Time of Change,* Vol. 2, pp. 208-243; Joan Parpart Zoeter, "USSR: Hard Currency Trade and Payments", in US Congress, Joint Economic Committee, *Soviet Economy in the 1980s: Problems and Prospects,* 97th Congress, 2d Session (Washington, D.C.: US Government Printing Office, 1983), Part 2, pp. 479-506; and Gregory Grossman and Ronald L. Solberg, *The Soviet Union's Hard-Currency Balance of Payments and Creditworthiness in 1985* (R-2956-USDP; Santa Monica, Calif.: Rand Corporation, 1983), ch. II.

41. Estimates of orders, compiled from incomplete press reports, can indicate only the general magnitude and trend, in contrast to the more precise information on deliveries furnished by trade statistics.
42. The interaction of some of these favourable and unfavourable factors in the 1970s is discussed in Zdenek Drabek, "External Disturbances and the Balance of Payments Adjustment in the Soviet Union", *Aussenwirtschaft,* Vol. 38 (1983), No. 2, pp. 173-194.
43. The Bank for International Settlements (BIS), *Manual on Statistics Compiled by International Organisations on Countries' External Indebtedness* (Basle, 1979), explains in detail what is included and excluded in estimates compiled by the BIS, International Bank for Reconstruction and Development (World Bank), International Monetary Fund, and OECD.
44. For earlier estimates, see, for example, Zoeter, "USSR: Hard Currency Trade", pp. 489-495, and Grossman and Solberg, *The Soviet Union's Hard-Currency Balance of Payments,* pp. 61-62.
45. The share of officially-supported credit in total Soviet gross debt may be higher than indicated in Table 9, according to other estimates. For the USSR plus the CMEA banks, a) trade-related bank claims under official insurance and b) officially-supported non-bank trade-related credits were estimated, respectively, at $5.6 billion and $12.3 billion out of total gross debt of $28.6 billion on 31st December 1982. The corresponding figures were $5.0 billion, $11.6 billion, and $27.1 billion on 31st December 1983. See OECD and BIS, *Statistics on External Indebtedness: Bank and Trade-Related Non-Bank External Claims on Individual Borrowing Countries and Territories at End-December 1982 and End-June 1983* (Paris and Basle, April 1984), p. 6, and *Statistics on External Indebtedness: Bank and Trade-Related Non-Bank External Claims on Individual Borrowing Countries and Territories at End-December 1983* (Paris and Basle, July 1984), p. 6. Separate figures for the USSR alone, without the CMEA banks, are not provided in the OECD-BIS statistics.
46. See, for instance, G. Fink, *An Assessment of European CMEA Countries' Hard Currency Debt* (Forschungsberichte No. 72; Vienna: Vienna Institute for Comparative Economic Studies, September 1981), p. 14.
47. For more detailed explanations and typologies of countertrade arrangements, see, for example, Pompiliu Verzariu, *Countertrade Practices in East Europe, the Soviet Union, and China: An Introductory Guide to Business* (US Department of Commerce; Washington, D.C.: US Government Printing Office, 1980); Organisation for Economic Co-operation and Development, *Recent Developments in Countertrade* (Paris: OECD, 1981); and US International Trade Commission, *Analysis of Recent Trends in US Countertrade* (Publication 1237; Washington, D.C., 1982).
48. For Soviet views, see, for instance, V. Ivashkin and V. Panchenko, "Vneshneekonomicheskie sviazi stran SEV s kapitalisticheskimi gosudarstvami" [Foreign economic ties of CMEA countries with capitalist states], *Voprosy ekonomiki,* 1979, No. 6, p. 117; G. Oleinik, "Sdelki na kompensatsionnoi osnove (opyt zarubezhnykh stran SEV)" [Transactions on a compensation basis (experience of foreign CMEA countries)], *Ekonomika i organizatsiia promyshlennogo proizvodstva,* 1980, No. 2, p. 182; and Taisiya Nirsha, "Compensation-Based Co-operation – A Mutually Advantageous Form of Economic Relations Between East and West", *Foreign Trade* (Moscow), 1981, No. 8, p. 37.
49. However, market fluctuations in prices of Soviet exports under compensation agreements are deemed a "great inconvenience" for the USSR, according to Sergei Ponomaraev and Vladimir Savin, "Mutually Beneficial Co-operation on a Compensation Basis", *Foreign Trade* (Moscow), 1983, No. 8, p. 37.
50. Unless otherwise noted, this composite picture is based on Barclay, "USSR: The Role of Compensation Agreements", pp. 465-472; Vladimir Savin, "Co-operation on a Compensation Basis", *Foreign Trade* (Moscow), 1980, No. 4, pp. 20-25; Verzariu, *Countertrade Practices,* pp. 44-46; and Jan Stankovsky, "Compensation in East-West Trade", *Soviet and Eastern European Foreign Trade,* Vol. XIII, No. 4 (Winter 1977-1978), pp. 3-24, translated from *Creditanstalt-Bankverein Wirtschaftsberichte,* 1977, No. 6, pp. 7-16.

51. OECD, *East-West Trade: Recent Developments in Countertrade,* pp. 42 and 75.

52. Franz-Lothar Altmann, "Countertrade with CMEA Countries: The West German Example", in *Economic and Financial Aspects of East-West Co-operation* (Vienna: Zentralsparkasse und Kommerzbank, 1979), p. 64.

53. Iu. Piskulov, "Vostok-Zapad: Vazhnyi rezerv sotrudnichestva" [East-West: A great reserve for co-operation], *Ekonomicheskaia gazeta,* 1983, No. 41, p. 21.

54. For a detailed account, see US Congress, Office of Technology Assessment, *Technology,* chs. II-V and VII-IX. Briefer appraisals include Gary Bertsch, Richard Cupitt, John R. McIntyre, and Miriam Steiner, "East-West Technology Transfer and Export Controls", *Osteuropa Wirtschaft,* Vol. 26, No. 2 (June 1981), pp. 116-136, and John P. Hardt and Kate S. Tomlinson, "The Potential Role of Western Policy toward Eastern Europe in East-West Trade", in Abraham S. Becker (ed.), *Economic Relations with the USSR: Issues for the Western Alliance* (Lexington, Mass.: Heath, 1983), pp. 111-128.

55. See Gary K. Bertsch, *East-West Strategic Trade, CoCom, and the Atlantic Alliance* (Paris: Atlantic Institute for International Affairs, 1983), pp. 33-52.

Chapter 3

MODES OF TRANSFER OF WESTERN TECHNOLOGY TO THE USSR

There are various possible classifications of modes of technology transfer that differ, for example, in:

1. The transfer mechanisms included;
2. The disaggregation of particular categories;
3. The emphasis on embodied versus disembodied technology;
4. The prominence accorded the nature and extent of Western firms' involvement in the transfer process; and
5. The treatment of relationships among modes.

The analysis of modes of transfer of Western technology to the USSR in this chapter considers in turn licence purchases (Section A), commodity imports (Section B), turnkey plants (Section C), industrial co-operation agreements (Section D), and other modes (Section E).

This organisational scheme reflects the form in which relevant information is available. It also facilitates comparisons among modes and an explanation of Soviet choices among them. Further, it illuminates the trend from more "passive" to more "active" transfer mechanisms. Among the more passive mechanisms are licences without know-how, and imports of machinery and equipment without technical services or in small one-time sales for reverse engineering. In contrast, more active mechanisms – entailing deeper foreign involvement – include technology packages like turnkey plants and long-term co-operative Western-Soviet relationships in production, technology updating, and trade.

Through purchases of foreign licences, the USSR acquires technology not available domestically and releases its R & D organisations for work on other products and processes. Also, in the long run licence purchases may save foreign exchange, when domestic production of a commodity replaces imports of it, or earn foreign exchange, when Soviet production based on the licence is exported. However, the magnitude of Soviet licence purchases is relatively low, in comparison with licence purchases of developed market economies and also in comparison with other modes of technology transfer to the USSR.

Soviet commodity imports can be deemed a vehicle of technology transfer when they embody – and thus provide benefits from – technology not otherwise available in the USSR. Although a technology gap may be responsible for imports of technologically advanced consumer goods, the analysis of technology transfer via commodity imports concentrates on producer goods to make other commodities. Such "embodied" technology transfer has occurred on a significant scale in Soviet imports of Western machinery, equipment, metal products, and chemicals, used primarily in the USSR's engineering and metalworking, chemical, and energy industries.

In turnkey projects, foreign firms supply whole production systems and thus may design facilities, supervise construction and installation work, train personnel, and aid in the start-up of production. Many such turnkey projects have been carried out in the Soviet chemical, ferrous metallurgy, machine building, and light and food industries.

Through industrial co-operation agreements (ICAs), the USSR obtains the long-term involvement of Western firms in Soviet production methods, product characteristics, quality control, and training of personnel. The most common kinds of Soviet-Western ICAs are for co-production and for specialisation. In the former, the partners specialise in producing components and then exchange them so each can make the same final product. In the latter, the partners specialise in making end-products and then exchange them so each has a full line for sale. More than a third of all Soviet ICAs are in the chemical industry.

The USSR makes an extensive effort to collect, analyse, abstract and translate, and disseminate scientific and technical information from many kinds of foreign publications. The USSR also secures such information from scholarly exchange programmes with Western countries. Finally, the USSR uses various illegal means to obtain some of the technology covered by Western export controls.

A. LICENCE PURCHASES

Through licence purchases the licensee obtains rights to make and sell specific products incorporating inventions and processes developed by the licensor. The licensee pays for the technology in a lump sum, through royalty payments based on resultant sales, or both. The contract states a time period for the life of the licence, often 5-10 years. The licence may provide for one-time disclosure of commercial secrets or a continuing flow of new information developed during the life of the contract. The licence may include restrictions on the level of output and sales to foreign markets.

There are several categories of licences:

1. "Pure" licences grant the right to use a patent (or trademark) without the transfer of other information, know-how, and technical documentation.
2. Other licences also include know-how, technical documentation, and technical assistance. The inclusion of know-how is often deemed much more critical than patent rights alone, because the licensee's goal is to reproduce the licensor's technological capabilities[1].
3. "Complex" licences link patents and know-how with deliveries of machinery and equipment.

Soviet licence purchases have several objectives.

One is to save convertible currency by purchasing licences to produce items in the USSR, rather than importing the corresponding goods at a higher foreign exchange cost.

Foreign licences free Soviet R & D resources for other purposes[2] – including those for which technology imports are not available (e.g. military programmes) and those in which the comparative disadvantage of Soviet, compared to Western, R & D is smaller.

Technology obtained through licence imports can greatly shorten the time to start new production. For instance, in the Briansk machine building factory several types of ship engines were assimilated into production in two years after the receipt of a foreign licence, whereas domestic development of the engines would have required 8-10 years[3].

Foreign licences also reduce the cost of introducing innovations. Soviet planning procedures specify that the estimated present value of the cost of production under a foreign licence (over the term of the licence) must be at least one-third less than that of production on the basis of domestic R & D[4]. A Soviet source reports that the "economic effect" (usually calculated in terms of cost reduction) of the utilisation of foreign licences over the period of the licence agreements was often 10 times the cost of the licences. Although the specific licences are not identified, this statement appears in connection with a discussion of foreign licences for construction materials, chemicals, electrohydraulic deck cranes, and polishing glass[5].

Furthermore, some of the Soviet production based on foreign licences is exported, including for convertible currencies, thus covering part of the cost of the licences[6].

However, there are a number of problems connected with Soviet purchases of Western licences.

First, domestic R & D capabilities may not be developed in the field covered by the licence, so that the Soviet R & D lag continues or grows.

Second, Western firms may be reluctant to supply the latest technology, preferring for competitive reasons to sell an aging technology instead. Also, the time required to negotiate licence purchases may be longer than that to arrange the purchase of the corresponding machinery and equipment from the Western firm.

After a licence has been bought, its implementation may be delayed. Technology for many kinds of machinery and equipment becomes obsolescent in 4-6 years. Hence, Soviet planners use an average normative period of 1.5 years for the time from the purchase of a licence to the beginning of production under it[7]. But, for reasons discussed in Chapter 2, frequently assimilation of new technology takes longer and its economic effect is smaller than expected[8]. Hence, Soviet specialists stress the importance of obtaining licences under agreements that provide for the Western supplier to participate in the "mastering" of the licence in Soviet industry and also to update the technology[9].

Statistical data on Soviet licence purchases from the West are scanty and often vague.

Western compilations, usually from reports in the Western specialised press, commonly identify the date, the type of technology (in greater or lesser detail), and the Western firm. But they do not ordinarily state the terms or value of the licence agreements. Such surveys show that Soviet licence purchases from the West have been concentrated in chemicals, electrical equipment, iron and steel, machine tools, and oil and gas. A variety of examples of Soviet licence purchases from the West are given in Annex Table A-2[10].

Soviet publications offer only very general quantitative information about licence imports from the West. Furthermore, these statements are not usually clear about the extent to which they cover only "pure" licences bought by Litsenzintorg, the specialised licence-trading foreign trade organisation, or also licences bought by other foreign trade organisations as part of package deals along with machinery and equipment.

For example, one Soviet source indicates that the number of Western licences bought by the USSR from an unspecified initial date before 1970 up to the end of 1976 was 2.4 times the 550 Soviet licences sold to the West during the period[11].

According to another Soviet source, in 1966-1973 on the basis of licences imported from capitalist countries the USSR assimilated over 40 new technological processes and began production of 80 types of machines, equipment, and materials[12].

In a book published in 1981, a Ministry of Foreign Trade official stated that "in the past two decades" the USSR purchased from the West 700 licences, of which 60 per cent were "straight" licences not connected with machinery and equipment purchases and the remainder were "associated" licences that were so linked. However, outlays for associated licences (without counting the related machinery and equipment) were triple those for straight licences. The cost of associated licences is often 10-25 per cent – but sometimes as much as 40-50 per cent – of the total package including machinery and equipment[13].

An earlier source put the share of "pure" (straight) licences in total licence payments at 22 per cent in 1968 and 34 per cent in 1972[14].

There are no satisfactory published data for the absolute value of Soviet licence purchases from, or royalty payments to, the West. By many rough assumptions, Hanson reached what he called a "guesstimate" of an order of magnitude of $55 million for the value of Soviet licence purchases in 1970[15]. However, the Soviet Union frequently buys licences on credit, sometimes repaying with exports of resultant products, and royalties rather than lump-sum payments have been the chief form of payment since 1970[16]. In an article asserting that Soviet licence purchases were far too low, in view of the superiority of foreign (i.e. Western) over Soviet technology in many fields, N.N. Smeliakov, a Deputy Minister of Foreign Trade of the USSR, stated that in 1980 the Soviet Union "spent on licences" $64 million. The coverage of this figure was not explained, but Smeliakov contrasted it with outlays on licences in 1979 of $300 million by the Federal Republic of Germany, $1.2 billion by Japan, and $700 million by the United States and the Netherlands[17].

Hence, a clear picture of the quantitative dimensions of Soviet licence imports from the West is not available. The amount spent on licence imports appears to be modest. However, the advantages explaining the relative importance (by value) of complex or associated, rather than pure or straight, licences are evident.

B. COMMODITY IMPORTS

Commodity imports can be regarded as a mode of technology transfer when they embody technologies not available in the recipient country. Such imports transfer not the methods of production themselves, but rather the products constituting the results of these technologies[18]. In some cases, a nation which possesses the relevant technology, but has not diffused it widely enough, may import part of its supply in order to supplement domestic production. However, not all imports can be explained by technological lags and embodied technology transfers. The level and composition of imports (and exports) also reflect inter-industry and intra-industry specialisation decisions made for comparative advantage and other reasons.

Although a technology gap may account for imports of "technologically- advanced" or "research-intensive" consumer goods, such as passenger cars and television sets, the analysis of technology transfer through commodity imports focuses on producer goods used to make other commodities. These producer goods include capital goods, like machinery, equipment, and instruments, and intermediate products, like metals and chemicals. Capital goods imports are usually quantitatively more important, and their role in speeding technological progress and expanding output is often clearer. But imports of intermediate products can also be very significant in increasing production, including for export.

Commodities embodying superior foreign technology may be imported only once or a few times in small lots so they can be analysed and copied, probably with some adaptation, for production in Soviet industry. Many examples of this practice were reported in Sutton's studies of the use of Western technology in Soviet industry before and after World War II[19].

But there are a number of limitations to the copying approach. First, it still requires scarce R & D resources, it is time-consuming, and it is not always successful. Second, even in successful cases, a technology gap continues or even grows. The Western model is usually obtained only after it has been in production for some years, and the imitation process takes additional years. Hence, a Soviet copy may enter production 4-10 years after the original Western version. Meanwhile, the Western firm is improving its product or process. Third, some technology cannot be effectively copied. Machinery and mechanical equipment can be bought, disassembled, and analysed with the aim of duplicating it. But such "reverse-engineering" is difficult for products like integrated circuits and not ordinarily feasible for process innovations. Finally, unauthorised copying may violate patent rights, and patent infringement suits could limit exports of imitative products to some foreign markets[20].

Compared to one-time purchases for copying, large-scale and continuing imports of machinery and equipment embodying superior Western technology involve larger outlays of convertible currency but are likely to have a faster, more certain, and greater impact on Soviet production and, potentially, exports.

For the study of technology transfer, it is desirable to distinguish the level of technology embodied in commodity imports. However, there are no standard, universally accepted definitions of "high technology", "advanced technology", "technology-based", or "research-intensive" commodities. Further, most trade data are not sufficiently detailed to permit tabulation of flows of goods according to whatever definitions are preferred. Hence, in practice, each compiler of commodity trade statistics relevant to technology transfer adopts definitions that appear reasonable on a combination of conceptual and empirical grounds[21].

Published official Soviet trade statistics do not provide the detailed commodity composition desirable for the appraisal of technology transfer, and the data furnished are often incomplete in regard to value, quantity, and identification of trading partners. Therefore, to examine the transfer of Western technology to the USSR through commodity trade, this study uses statistics specially prepared for this purpose for the OECD. The data in Tables 10-17 cover relevant exports of OECD countries to the USSR. They are based on larger tables in the Annex which present the OECD statistical tabulations in greater detail and for additional years. These OECD statistics give the most complete picture available, although they too necessarily have various limitations in coverage and disaggregation[22].

Data on OECD countries' exports to the USSR of industrial goods by branch of origin in 1970, 1976, and 1982 are given in Table 10. They show the dominance of machinery, equipment, and metal products; metallurgy; and chemicals – with respective shares in 1982 of 28.2, 23.4, and 13.5 per cent. Within the category of machinery, equipment, and metal products, the most important sub-branches are metal-working and wood-working machinery (representing about one-fifth of the branch total), ships and motor vehicles, construction and mining machinery, and scientific, measuring, and control equipment. Metallurgy was among the fastest-growing branches. The chief component of these exports is iron and steel, primarily pipe. In the chemicals category, basic industrial chemicals accounted for about two-fifths of the total in 1982. Next in importance, with about one-fourth of the chemicals total in 1982, are synthetic resins, plastics, and artificial fibres.

A significant part of the growth in the nominal value of OECD industrial goods exports to the USSR since 1970 has been due to price increases. Table 11 presents data in 1970 US dollars for these exports in 1970, 1976, and 1981. From 1970 to 1976, total industrial goods exports rose 360 per cent in current dollars, but only 138 per cent in constant dollars. From 1970 to 1981, the nominal value increased sixfold (see Annex Table A-3), but the deflated value only tripled.

Table 10. **OECD industrial goods exports to the USSR, by branch of origin, 1970, 1976, and 1982[a]**

Branch of Origin	1970		1976			1982		
	Value (Millions of US dollars)	% of total	Value (Millions of US dollars)	% of total	Index (1970 = 100)	Value (Millions of US dollars)	% of total	Index (1970 = 100)
Products of mining	9.1	0.4	28.9	0.3	318	199.6	1.1	2 193
Coke and manufactured gas	0.0[b]	0.0	0.0	0.0	100	0.0	0.0	100
Energy	0.0	0.0	0.0	0.0	100	0.0	0.0	100
Food processing	91.7	3.8	391.5	3.5	427	1 745.6	9.4	1 904
Textiles	184.0	7.6	371.0	3.3	202	581.1	3.1	316
Clothing	99.0	4.1	133.9	1.2	135	319.9	1.7	323
Leather, shoes, and furs	59.8	2.5	118.3	1.1	198	275.4	1.5	460
Paper, pulp, and processed wood	193.2	7.9	503.3	4.5	261	1 112.3	6.0	576
Products of printing industry	7.6	0.3	23.8	0.2	313	44.5	0.2	586
Glass and china	6.9	0.3	18.0	0.2	261	28.8	0.2	417
Chemicals	292.1	12.0	1 004.4	9.0	345	2 503.5	13.5	857
Other non-metallic mineral products	5.9	0.2	52.4	0.5	888	73.6	0.4	1 247
Metallurgy	330.4	13.6	2 867.3	25.6	868	4 346.5	23.4	1 316
Machinery, equipment, and metal products	888.9	36.5	4 001.6	35.8	450	5 238.7	28.2	589
Other industrial commodities	263.6	10.8	1 666.0	14.9	632	2 105.9	11.3	799
Total	2 432.3	100.0	11 180.2	100.0	460	18 575.4	100.0	764

a) The classification corresponds to the USSR branch of origin classification. Components may not add to totals because of rounding.
b) Less than 0.1 million US dollars.
Source: Annex Table A-3.

Table 11. **Deflated value of OECD industrial goods exports to the USSR, by branch of origin, 1970, 1976 and 1981**[a]

Branch of origin	1970		1976			1981		
	Value (millions of 1970 US dollars)	% of total	Value (millions of 1970 US dollars)	% of total	Index (1970=100)	Value (millions of 1970 US dollars)	% of total	Index (1970=100)
Products of mining	9.1	0.4	20.0	0.3	220	111.0	1.4	1 209
Coke and manufactured gas	0.0[b]	0.0	0.0	0.0	100	0.0	0.0	100
Energy	0.0	0.0	0.0	0.0	100	0.0	0.0	100
Food processing	91.7	3.8	273.7	4.7	298	1 200.1	15.3	1 309
Textiles	184.0	7.6	210.5	3.6	114	364.6	4.6	198
Clothing	99.0	4.1	66.9	1.2	68	160.3	2.0	162
Leather, shoes, and furs	59.8	2.5	92.0	1.6	154	131.0	1.7	219
Paper, pulp, and processed wood	193.2	7.9	296.0	5.1	153	600.0	7.7	311
Products of printing industry	7.6	0.3	10.7	0.2	141	16.0	0.2	211
Glass and china	6.9	0.3	10.8	0.2	157	9.4	0.1	136
Chemicals	292.1	12.0	711.1	12.3	243	1 263.6	16.1	433
Other non-metallic mineral products	5.9	0.2	31.5	0.5	534	25.6	0.3	434
Metallurgy	330.4	13.6	1 464.5	25.2	443	1 742.6	22.2	527
Machinery, equipment, and metal products	888.9	36.5	1 894.8	32.7	213	1 533.0	19.6	172
Other industrial commodities	263.6	10.8	717.6	12.4	272	683.6	8.7	259
Total	2 432.3	100.0	5 800.1	100.0	238	7 840.5	100.0	322

a) The classification corresponds to the USSR branch of origin classification. Components may not add to totals because of rounding.
b) Less than 0.1 million US dollars.
Source: Annex Table A-4.

Table 12 contains data on OECD countries' exports to the USSR of capital goods, by type of product. The largest category (also among the most rapidly growing) is liquid fuel, gas, and water distribution – mainly tubes and pipes – representing about a fifth of the total in 1976 and almost a third in 1982. Next in importance are transport equipment (primarily ships); construction and mining machinery; and engineering, welding, and metallurgical equipment – chiefly machine tools. Of the total value of machine tool sales, metal-cutting tools accounted for about half; metal-forming tools, about one-third; and transfer lines, the remainder. Within the metal-cutting tool category, about 80 per cent (by value) were conventional tools and only about 20 per cent numerically-controlled[23].

The distribution of these OECD capital goods exports to the USSR by end use is given in current US dollars in Table 13 and in constant US dollars in Table 14. According to Table 13, in 1982 Soviet industry received almost half of the total, and oil and gas distribution about a fourth (though its share rose significantly over the period). Transportation was third in importance, with about a seventh of the total.

The data for the Industry row in Tables 13 and 14 are disaggregated in Tables 15 and 16, respectively, to show the distribution by branch within Soviet industry of OECD countries' exports of capital goods to the USSR. As Table 15 indicates, in 1982 mining and fuel extraction machinery was the most important category, with 17.2 per cent of the total (compared to 4.8 per cent in 1976). Close behind was the engineering and metalworking industry, with 16.0 per cent of the total in 1982 (compared to 19.6 per cent in 1976). The non-attributable category in these tables is rather large, about one-third of the total. It consists chiefly of electricity distribution equipment, machinery and mechanical appliances, and metal structural parts which could not be assigned to particular branches. If such attribution could be made, it might alter significantly the shares of particular branches, for example, raising the figures for the chemical industry[24].

As noted earlier, technology transfer may occur through imports of intermediate products as well as of machinery and equipment. Table 17 presents data on OECD countries' exports of technology-based intermediate goods to the USSR by type of product. The most important single category is iron and steel, representing two-fifths of the total in 1976 and one-third in 1982, and showing a 15-fold increase (in current prices) from 1970 to 1982. Other significant product categories in the table are:

1. Organic and inorganic chemical elements; and
2. Chemicals and plastic materials.

These two categories together have amounted to between one-fifth and one-fourth of the total. Finally, shipments of parts and accessories of machinery and equipment have grown along with the sales of the capital goods for which they are used.

The data furnished in Tables 10-17 thus show the role of such Western exports as machinery, equipment, metal products, and chemicals in the transfer of technology to the USSR, particularly its energy, chemical, engineering, and mining industries.

Examination of the paired tables in current US dollars and constant US dollars (Tables 10 and 11, 13 and 14, and 15 and 16 – and the corresponding Annex tables) reveals that an important factor in the growth in the value of these exports was the increase in world market prices after 1970, and especially after 1973. Although the USSR benefited from markedly higher prices for its oil exports, it also had to pay more for Western capital goods and intermediate products. Hence, the real growth in Soviet imports of commodities of both types was much less than the nominal growth.

However, the impact of inflation varied by product group. For example, from 1970 to 1981 the OECD countries' total exports of capital goods to the USSR rose 398 per cent in

Table 12. **OECD capital goods exports to the USSR, by type of product, 1970, 1976 and 1982**[a]

Type of product	1970		1976			1982		
	Value (Millions of US dollars)	% of total	Value (Millions of US dollars)	% of total	Index (1970=100)	Value (Millions of US dollars)	% of total	Index (1970=100)
Stationary power plants and water engineering	7.0	0.6	134.3	2.4	1 919	186.9	2.6	2 670
Electric power distribution	37.9	3.4	140.8	2.5	372	219.7	3.0	580
Liquid fuel, gas, and water distribution	181.8	16.1	1 340.9	23.7	738	2 170.7	29.9	1 194
Transport equipment	195.0	17.3	830.9	14.7	426	1 059.5	14.6	543
Agricultural equipment	25.5	2.3	180.7	3.2	709	116.2	1.6	456
Construction and mining machinery	30.9	2.7	207.9	3.7	673	595.7	8.2	1 928
Engineering, welding, and metallurgical equipment (excluding furnaces)	191.8	17.0	658.0	11.6	343	549.0	7.6	286
Industrial and laboratory furnaces and gas generators	53.4	4.7	153.9	2.7	288	122.2	1.7	229
Machine and hand tools for working minerals, wood, plastics, etc.	13.2	1.2	43.0	0.8	326	62.3	0.9	472
Other electrical equipment	12.5	1.1	23.5	0.4	188	24.2	0.3	194
Electronics and telecommunications	38.0	3.4	87.1	1.5	229	137.3	1.9	361
Machinery for special industries	85.7	7.6	346.5	6.1	404	348.7	4.8	407
Mechanical handling equipment and storage tanks	46.5	4.1	260.3	4.6	560	492.0	6.8	1 058
Office machines	1.0	0.1	3.1	0.1	310	2.9	0.0	290
Medical apparatus, instruments, and furniture	5.5	0.5	32.6	0.6	593	51.3	0.7	933
Heating and cooling of buildings and vehicles	2.1	0.2	11.8	0.2	562	31.2	0.4	1 486
Measuring, controlling, and scientific instruments	52.6	4.7	156.5	2.8	298	227.8	3.1	433
Finished structural parts and structures	8.8	0.8	149.2	2.6	170	242.5	3.3	2 756
Other capital equipment	137.1	12.2	902.4	15.9	658	621.5	8.6	453
Total	1 126.4	100.0	5 663.4	100.0	503	7 261.5	100.0	645

a) Components may not add to totals because of rounding.
Source: Annex Table A-5.

Table 13. **OECD capital goods exports to the USSR, by end use, 1970, 1976, and 1982** [a]

End Use	1970		1976			1982		
	Value (Millions of US dollars)	% of total	Value (Millions of US dollars)	% of total	Index (1970=100)	Value (Millions of US dollars)	% of total	Index (1970=100)
Industry	617.0	54.8	2 946.1	52.0	477	3 269.8	45.0	530
Oil and gas distribution	175.6	15.6	1 279.3	22.6	729	2 099.2	28.9	1 195
Agriculture	25.5	2.3	180.7	3.2	709	116.2	1.6	456
Construction	0.0 [b]	0.0	0.0	0.0	100	0.2	0.0	[c]
Domestic trade	0.0	0.0	0.1	0.0	[c]	0.0	0.0	100
Transport	194.9	17.3	812.7	14.3	417	1 030.7	14.2	529
Communications	17.0	1.5	28.1	0.5	165	63.0	0.9	371
Administration and management	19.6	1.7	55.4	1.0	283	58.9	0.8	300
Public health	5.5	0.5	32.6	0.6	593	51.3	0.7	933
Non-attributable	71.6	6.4	329.8	5.8	461	578.2	8.0	808
Total	1 126.8	100.0	5 664.8	100.0	503	7 267.4	100.0	645

a) Components may not add to totals because of rounding.
b) Less than 0.1 million US dollars.
c) Not applicable, as an index number cannot be computed with a zero denominator.
Source: Annex Table A-6.

Table 14. **Deflated value of OECD capital goods exports to the USSR, by end use, 1970, 1976, and 1981**[a]

End Use	1970		1976			1981		
	Value (Millions of 1970 dollars)	% of total	Value (millions of 1970 dollars)	% of total	Index (1970=100)	Value (millions of 1970 dollars)	% of total	Index (1970=100)
Industry	617.0	54.8	1 353.1	50.2	219	897.5	41.9	145
Oil and gas distribution	175.6	15.6	644.2	23.9	367	673.6	31.4	384
Agriculture	25.5	2.3	85.5	3.2	335	56.7	2.6	222
Construction	0.0[b]	0.0	0.0	0.0	100	0.0	0.0	100
Domestic trade	0.0	0.0	0.1	0.0	[c]	0.0	0.0	100
Transport	194.9	17.3	389.1	14.4	200	246.9	11.5	127
Communications	17.0	1.5	18.7	0.7	110	34.4	1.6	202
Administration and management	19.6	1.7	35.2	1.3	180	58.6	2.7	299
Public health	5.5	0.5	16.5	0.6	300	22.1	1.0	402
Non-attributable	71.6	6.4	152.9	5.7	214	152.4	7.1	213
Total	1 126.8	100.0	2 695.2	100.0	239	2 142.2	100.0	190

a) Components may not add to totals because of rounding.
b) Less than 0.1 million US dollars.
c) Not applicable, as an index number cannot be computed with a zero denominator.
Source: Annex Table A-7.

Table 15. **OECD industrial capital goods exports to the USSR, by industrial end-use branch, 1970, 1976, and 1982**[a]

End-Use Branch	1970		1976			1982		
	Value (Millions of US dollars)	% of total	Value (Millions of US dollars)	% of total	Index (1970=100)	Value (Millions of US dollars)	% of total	Index (1970=100)
Mining and fuel extraction machinery	26.3	4.3	140.3	4.8	533	562.1	17.2	2 137
Raw materials processing industry	5.8	0.9	70.4	2.4	1 214	41.8	1.3	721
Energy-power plants	6.9	1.1	130.4	4.4	1 890	186.2	5.7	2 699
Food processing and tobacco industry	14.7	2.4	79.2	2.7	539	98.0	3.0	667
Textile industry	21.0	3.4	150.9	5.1	719	126.7	3.9	603
Clothing industry	2.7	0.4	10.0	0.3	370	18.8	0.6	696
Leather, shoe, and fur industry	5.5	0.9	13.0	0.4	236	8.2	0.2	149
Paper, pulp, and wood-processing industry	34.4	5.6	101.3	3.4	294	96.8	3.0	282
Printing industry	15.2	2.5	18.5	0.6	122	37.3	1.1	245
Glass and china industries	6.1	1.0	19.2	0.7	315	20.6	0.6	338
Engineering and metalworking industry	173.8	28.2	576.9	19.6	332	524.3	16.0	302
Metallurgy	63.5	10.3	158.3	5.4	249	123.8	3.7	195
Coke and gas industry	7.8	1.3	69.6	2.4	892	15.6	0.5	200
Chemicals and construction materials	45.8	7.4	471.5	16.0	1 029	313.1	9.6	684
Non-attributable	187.7	30.4	936.6	31.8	499	1 096.6	33.5	584
Total	617.0	100.0	2 946.1	100.0	477	3 269.8	100.0	530

a) Components may not add to totals because of rounding.
Source: Annex Table A-8.

Table 16. **Deflated value of OECD industrial capital goods exports to the USSR, by industrial end-use branch, 1970, 1976, and 1981[a]**

End-Use Branch	1970		1976			1981		
	Value (millions of 1970 US dollars)	% of total	Value (millions of 1970 US dollars)	% of total	Index (1970=100)	Value (millions of 1970 US dollars)	% of total	Index (1970=100)
Mining and fuel extraction machinery	26.3	4.3	59.7	4.4	227	107.5	12.0	409
Raw materials processing industry	5.8	0.9	32.9	2.4	567	14.4	1.6	248
Energy-power plants	6.9	1.1	62.3	4.6	903	26.2	2.9	380
Food processing and tobacco industry	14.7	2.4	36.7	2.7	250	31.1	3.5	212
Textile industry	21.0	3.4	72.2	5.3	344	36.7	4.1	175
Clothing industry	2.7	0.4	5.2	0.4	193	5.2	0.6	193
Leather, shoe, and fur industry	5.5	0.9	6.2	0.5	113	6.5	0.7	118
Paper, pulp, and wood-processing industry	34.4	5.6	45.5	3.4	132	24.3	2.7	71
Printing industry	15.2	2.5	8.2	0.6	54	15.1	1.7	99
Glass and china industries	6.1	1.0	9.1	0.7	149	8.6	1.0	141
Engineering and metalworking industry	173.8	28.2	254.7	18.8	147	147.4	16.4	85
Metallurgy	63.5	10.3	76.4	5.6	120	37.4	4.2	59
Coke and gas industry	7.8	1.3	31.1	2.3	399	5.2	0.6	67
Chemicals and construction materials	45.8	7.4	208.8	15.4	456	73.8	8.2	161
Non-attributable	187.7	30.4	444.0	32.8	237	358.0	39.9	191
Total	617.0	100.0	1 353.1	100.0	219	897.5	100.0	145

a) Components may not add to totals because of rounding.
Source: Annex Table A-9.

Table 17. **OECD exports of technology-based intermediate goods to the USSR, by type of product, 1970, 1976 and 1982**[a]

Type of product	1970		1976			1982		
	Value (Millions of 1970 US dollars)	% of total	Value (millions of 1970 US dollars)	% of total	Index (1970=100)	Value (millions of 1970 US dollars)	% of total	Index (1970=100)
Parts and accessories of machinery and equipment	76.8	9.5	472.1	11.9	615	827.4	12.6	1 077
Parts and accessories of transport equipment	33.9	4.2	154.5	3.9	456	386.7	5.9	1 141
Paper manufactures	130.2	16.0	354.4	8.9	272	584.2	8.9	449
Textile manufactures	114.9	14.2	282.3	7.1	246	419.7	6.4	365
Synthetic rubber	3.6	0.4	30.4	0.8	844	52.8	0.8	1 467
Synthetic fibres	28.4	3.5	27.8	0.7	98	41.9	0.6	148
Special manufactures of leather, rubber, wood, glass, and minerals	0.7	0.1	15.5	0.4	2 214	16.5	0.2	2 357
Miscellaneous mineral manufactures	3.9	0.5	10.0	0.3	256	16.7	0.2	428
Iron and steel	137.1	16.9	1 697.9	42.7	1 238	2 197.7	33.4	1 603
Non-ferrous metals	32.0	3.9	72.1	1.8	225	202.3	3.1	632
Organic and inorganic chemical elements	100.7	12.4	408.8	10.3	406	847.7	12.9	842
Final chemical manufactures	26.6	3.3	59.5	1.5	224	138.1	2.1	519
Chemicals and plastic materials	120.8	14.9	379.0	9.5	314	838.2	12.7	694
Other intermediate goods	2.4	0.3	8.2	0.2	342	11.5	0.2	479
Total	811.8	100.0	3 972.4	100.0	489	6 581.3	100.0	811

a) Components may not add to totals because of rounding.
Source: Annex Table A-10.

nominal terms, as a result of real growth of 90 per cent and price increases of 162 per cent. Inflation was less severe for capital goods for oil and gas distribution, for which the corresponding percentage figures are 767, 284 and 126. In contrast, prices rose faster for mining and fuel extraction machinery, for which real growth was 309 per cent and price inflation was 204 per cent, yielding an increase of 1 143 per cent in current dollars. (For other such comparisons, see the disaggregated annual statistics in Annex Tables 3-9.)

C. TURNKEY PROJECTS

Turnkey projects are a much more comprehensive and potentially more effective mode of technology transfer than separate commodity imports or licence purchases. In turnkey projects, foreign firms undertake to supply whole production systems and thus may provide feasibility studies, design of facilities, supervision of construction, delivery and installation of plant and equipment, licences, training for Soviet engineers and technicians in the USSR and abroad, and assistance in the commissioning of facilities and the start-up of production. Often there is a subsequent continuing relationship with the foreign firms through technical co-operation agreements and contracts for servicing machinery and equipment, for updating the technology initially puchased, for marketing exports, or for some combination of these.

Although foreign participation in the turnkey project is critical, some of the equipment and materials and most of the labour for the project are of Soviet origin. Also, often a number of foreign firms, perhaps from different countries, are involved in supplying elements of the project – rather than a single foreign firm providing the entire foreign participation. Sometimes a foreign firm or consortium acts as general contractor for the project, but this responsibility is usually performed by a Soviet organisation.

From the Soviet viewpoint, compared to commodity imports or licence purchases, turnkey projects have the advantages of also obtaining a variety of design, organisational, and production expertise that shortens the time required for the commissioning and successful operation of complex new facilities[25].

There are no regularly published official Soviet (or Western) statistical tabulations on turnkey projects. However, trade statistics capture the portion of a turnkey project involving commodity imports of machinery, equipment, and intermediate products, along with the value of complementary technical assistance included in the price charged for the commodity.

Hundreds of Soviet turnkey projects – of many kinds, in various industrial branches, of different sizes, and in many locations – have been identified from press reports[26]. However, sometimes the value of contracts is not known, and subsequent progress in the implementation of contracts is not regularly reported.

Contracts with foreign firms for turnkey projects in the USSR often are for hundreds of millions of dollars, and they sometimes exceed a billion dollars. Turnkey projects have been used in the Soviet chemical, ferrous metallurgy, motor vehicle, machine building, and light and food industries. Many turnkey projects – particularly those in chemicals and natural resource development – involve compensation (product payback) arrangements.

D. INDUSTRIAL CO-OPERATION

Industrial co-operation agreements (ICAs) establish a contractual relationship between a Soviet organisation and a Western partner under which they pool certain assets and jointly co-ordinate their use in mutual pursuit of complementary objectives. The pooled assets may be tangible (plant or equipment) or intangible (patents or know-how). They may be physically transferred between partners or remain in place. The co-operation may cover production, investment, marketing, financing, and/or research and development. Payments between partners may be in money or in kind[27].

Sometimes licensing and turnkey projects, considered separately in this study in Sections A and C above, are done under ICAs. Also, ICAs may involve countertrade, discussed in Chapter 2.

From the Soviet viewpoint, the advantage of ICAs – compared, for instance, with commodity imports or straight licence purchases – is the deeper and continuing contribution of the Western partner to the development of Soviet production. The foreign firm has a long-term involvement in the Soviet firm's production methods and costs, product characteristics, and quality control. Soviet-Western co-operation entails frequent and detailed communication; the transfer of specific proprietary information from the Western firm, including data on changes in product and process design; and exchanges of managerial, engineering, and technical personnel. From this co-operation the Soviet side expects a shorter start-up time for a new plant (often 1-2 years earlier), lower costs, a quality level close to that of the Western partner's own plants, and the possibility of exporting to the world market as well as to other CMEA countries[28].

Some Soviet scientific-technical agreements – especially those of the USSR State Committee for Science and Technology (SCST) – with the West are not specific ICAs but rather general framework "co-operation agreements". They identify areas of mutual interest and provide an organisational aegis for subsequent exchanges of people and information. However, the SCST has no authority to sign contracts for licences or commodity imports. Hence, these SCST "umbrella" (or "protocol") agreements must be implemented by further specific agreements linking the Western partners with the relevant Soviet branch ministry and the appropriate Soviet foreign trade organisation[29]. Thus, although there were reported to be over 280 such co-operation agreements between the SCST and Western firms in 1980[30], their impact on the transfer of technology depends on the extent to which they were complemented by specific follow-up ICAs.

The wide range of framework and implementing co-operation agreements is illustrated by the many examples of Soviet-Western agreements in Annex Table A-11. The table shows the distribution of over 200 agreements by branch of the economy, subject, Western firm, and Soviet organisation.

A quantitative evaluation of Soviet-Western ICAs is difficult because the available published information has been compiled chiefly from press reports and sample surveys of private Western firms. These estimates commonly report *the number of agreements signed* – sometimes including SCST umbrella agreements as well as operational ICAs, and unimplemented and expired as well as active ICAs. But the estimates do not tell the *value of contracts* signed under ICAs, progress in their actual *implementation,* or their *effect* on the volume and composition of the partners' output and sales or on international flows of goods and services.

Results of a 1983 study of a sample of 218 active Soviet-Western ICAs are presented in Tables 18-20. Although these data refer only to the percentage shares of the total number of contracts (without regard to their value), they give some idea of the pattern of Soviet-Western ICAs.

Table 18 shows the distribution of these Soviet-Western ICAs by their principal purpose, although some ICAs have more than one purpose. Co-production and specialisation were clearly the most important category, accounting for over half of all the agreements in the sample. In co-production, the partners – usually on the basis of some shared technology and agreed specifications – specialise in producing components and then exchange them so each can make the same final product for sale in its own market area. In product specialisation, the partners specialise in making end-products and then exchange them so each has a full line for sale in its market area. The Western firm usually provides the technology for the Soviet operation, which uses Soviet labour and materials[31]. However, joint research and development have occurred in the chemical, machine building, electronics, and electrical equipment industries[32]. Most of the co-production and specialisation agreements give the Western partner unique or principal access to Western markets, and the Soviet partners's market area is the CMEA countries and sometimes Third World nations[33].

The delivery of plant or equipment for turnkey projects is the second most important category in Table 18, accounting for a little more than one-fourth of the total number of agreements. Soviet-Western joint ventures, located in Western countries and organised to market Soviet exports for convertible currencies, were the subject of 7.3 per cent of the agreements. Another 6.9 per cent of the ICAs was for joint tendering or joint projects, involving Soviet-Western co-operation in construction, marketing, and finance in less developed countries.

By branch of industry (see Table 19), the chemical industry (including pharmaceuticals) dominated, with more than one-third of all the ICAs. Mechanical engineering, machine tools, electrical equipment, and transport equipment together accounted for almost one-third of the agreements.

As Table 20 shows, certain types of ICAs are concentrated in particular industries. For example, two-thirds of all the ICAs for plant and equipment deliveries were in the chemical industry. It also had the largest number of co-production and specialisation agreements – followed by mechanical engineering, transport equipment, and electronics.

Although the data in Tables 18-20 refer only to the percentage of the total number (not value) of the 218 contracts in the sample, the figures indicate the importance of co-production and specialisation and turnkey projects among the types of ICAs, and the concentration of ICAs in selected branches, including chemicals, machinery and equipment, electronics, and metallurgy.

Table 18. **Sample of 218 Soviet industrial co-operation agreements with Western firms, September 1983**

Distribution by Type of Agreement

Type of Agreement	Per cent of Total
Licensing	4.1
Delivery of plant or equipment	27.1
Co-production and specialisation	54.6
Sub-contracting	0.0
Joint ventures	7.3
Joint tendering or joint projects	6.9
Total	100.0

Source: United Nations, Economic Commission for Europe, *Statistical Survey of Recent Trends in Industrial Co-operation* (TRADE/R. 468; Geneva, 10th November 1983), Table 3, p. 9.

Table 19. **Sample of 218 Soviet industrial co-operation agreements with Western firms, September 1983**

Distribution by Branch of Industry

Branch of Industry	Per cent of Total
Chemical industry [a]	36.7
Metallurgy [b]	10.1
Transport equipment [c]	6.9
Machines tools	5.5
Mechanical engineering [d]	14.7
Electronics [e]	6.0
Electrical equipment [f]	2.7
Food and agriculture [g]	4.1
Light industry [h]	6.4
Other branches [i]	6.9
Total	100.0

a) Including pharmaceuticals.
b) Including mining.
c) Including aircraft, automobiles, trucks, tractors (even for agriculture), rolling stock, earth-moving equipment, and diesel engines (even stationary).
d) All other non-electrical engineering.
e) Including computers and other office equipment, radio and television sets, and communications equipment.
f) Including electric locomotives and household appliances.
g) Including beverages.
h) Including textiles, footwear, rubber, glass, furniture, and other consumer goods.
i) Including construction, hotel management, and tourism.
Source: United Nations, Economic Commission for Europe, *Statistical Survey of Recent Trends in Industrial Co-operation* (TRADE/R. 468; Geneva, 10th November 1983), Table 2, p. 8.

Table 20. **Sample of 218 Soviet industrial co-operation agreements with Western firms, September 1983**

Distribution by Type of Agreement in Branch of Industry [a], Per Cent

Branch of Industry	Licensing	Delivery of Plant or Equipment	Co-production and Specialisation	Sub-contracting	Joint Ventures	Joint Tendering or Joint Projects
Chemical industry [b]	33.3	67.8	26.1	0	18.7	20.0
Metallurgy [c]	11.1	15.2	6.7	0	12.5	13.3
Transport equipment [d]	22.2	0	10.9	0	0	0
Machine tools	0	1.7	7.6	0	12.5	0
Mechanical engineering [e]	0	8.5	16.8	0	25.0	20.0
Electronics [f]	11.1	1.7	9.2	0	0	0
Electrical equipment [g]	0	0	5.0	0	0	0
Food and agriculture [h]	0	0	5.0	0	12.5	6.7
Light industry [i]	22.2	5.1	5.9	0	6.2	6.7
Other branches [j]	0	0	6.7	0	12.5	33.3
Total, all branches	100.0	100.0	100.0	0	100.0	100.0

a) Component may not add to totals because of rounding.
b) Including pharmaceuticals.
c) Including mining.
d) Including aircraft, automobiles, trucks, tractors (even for agriculture), rolling stock, earth-moving equipment, and diesel engines (even stationary).
e) All other non-electrical engineering.
f) Including computers and other office equipment, radio and television sets, and communications equipment.
g) Including electric locomotives and household appliances.
h) Including beverages.
i) Including textiles, footwear, rubber, glass, furniture, and other consumer goods.
j) Including construction, hotel management, and tourism.
Source: United Nations, Economic Commission for Europe, *Statistical Survey of Recent Trends in Industrial Co-operation* (TRADE/R. 468; Geneva, 10th November 1983), Table 5, p. 11.

E. OTHER MODES

The USSR also obtains Western technology through a variety of "non-commercial" channels, including publications and trade shows, scientific exchange programmes, and illegal methods.

The Soviet Union has an extensive network of organisations for the collection, analysis, abstracting and translation, and dissemination of scientific and technical information in foreign publications, including textbooks, reference books, journals, technical standards manuals, and patent descriptions. In addition, Soviet specialists attend many foreign industrial and trade exhibitions and scientific meetings[34].

There are four main Soviet organisations concerned with the acquisition and distribution of documentary information on foreign technology[35]. The All-Union Institute of Scientific and Technical Information obtains, indexes, and abstracts a wide variety of foreign scientific and technical literature. The All-Union Scientific Research Institute of Interbranch Information handles classified and military-related materials. The Central Scientific Research Institute of Patent Information and Technical-Economic Studies collects and indexes foreign (and domestic) patent literature. The All-Union Scientific Research Institute of Technical Information, Classification, and Coding is responsible for information on technical norms and industrial standards.

These four institutes have an ambitious programme to collect and disseminate information on foreign technology. However, the collection effort is constrained by limited allocations of convertible currency to purchase books and journals abroad. Also, there are lags in the indexing, abstracting, and translation of the materials acquired. Finally, the publications containing the results of the institutes' work are issued in relatively small editions, compared to the demand for them by R & D personnel[36].

Various Soviet R & D and economic organisations follow industrial trends in developed market economies through the study of Western patent data, which cover most of the important areas of industrial technology. Patent information is available in machine-readable form from the International Patent Documentation Center (in Vienna), to which the USSR belongs. The information can easily be manipulated for patent searches by technological area, patent family, or patent-holding firm. Such patent information is used in Soviet decisions about purchases of foreign licences or goods and also in Soviet decisions about domestic R & D programmes[37].

Compared to other methods of acquiring Western technology, documentary collection has both advantages and disadvantages from the Soviet viewpoint[38]. On the one hand, the outlay of convertible currency is relatively small, and information can be obtained from countries with which the USSR does not have cordial political relations and from firms with which it does not have established commercial ties. On the other hand, much of the information secured through publications concerns fundamental research rather than applied research and development, refers to older rather than the latest technology, or is intended for sales promotion rather than production.

The USSR has scholarly exchange programmes with many Western countries. Although these programmes are arranged under intergovernmental agreements, the Western partners at the operating level are often national academies of sciences, cultural relations bodies, scientific associations, or universities. Under some programmes, individual Soviet scientists participate in symposia, make short visits to Western scientific institutions, or spend as much as a year in sustained research at a particular facility. Under other programmes, joint

Soviet-Western working groups present their respective research results at periodic conferences and may conduct some research jointly.

Such exchange programmes are attractive to the Soviet Union for several reasons. They enhance the training and qualifications of Soviet scientists. They are a vehicle to secure information about current research and development at the frontier of knowledge (in contrast to the older data in published materials)[39]. Finally, a considerable share of the Soviet scientists sent to the West under exchange programmes are in fields in which Soviet technology lags behind the West, such as computers, optics, and electronics[40].

However, a recent study of Soviet-United States scientific and technical exchange programmes, for example, concluded that they had little direct impact on the flow of advanced technology in the USSR, chiefly because most of the activities under these programmes involved basic or fundamental research, rather than technology with direct production applications (in contrast to industrial co-operation agreements)[41].

Finally, the USSR has been able through several methods to obtain some of the technology covered by the Western export controls discussed in Chapter 2, Section C. A legal method is the establishment of firms chartered as local companies in the United States or a West European country but owned by the Soviet Union (or an East European nation). Such firms can legally purchase controlled technology and study it in the country of origin, although they cannot legally export the goods or data without an export licence. Some Western technology is secured through industrial espionage, according to press reports. Another illegal method involves the use of intermediary firms that procure valid Western export licences for an approved end-use in an authorised country of destination and then reship the goods to the USSR[42].

However, most Soviet acquisition of Westen technology is by legal commercial means. In the 1970s, compared to the 1960s, the USSR placed more emphasis on "active" technology transfer mechanisms, like turnkey projects and industrial co-operation agreements, as distinct from "passive" modes like straight licences and commodity imports. This trend reflects the Soviet belief that the more active modes can significantly facilitate the assimilation of Western technology and thus increase its impact on the Soviet economy.

NOTES AND REFERENCES

1. Farok J. Contractor, *International Technology Licensing* (Lexington, Mass.: D.C. Heath and Co., 1981), pp. 33-34.
2. A.V. Boichenko, "Nauchno-tekhnicheskie sviazi stran sotsialisticheskogo sodruzhestva s kapitalisticheskimi gosudarstvami" [Scientific-technical ties of countries of the socialist commonwealth with capitalist states], *Vestnik Moskovskogo universiteta, Seriia 6: Ekonomika,* 1981, No. 2, p. 92.
3. Iu. Naido and S. Simanovskii, "Uchastie stran SEV v mirovoi litsenzionnoi torgovle" [Participation of CMEA countries in the world licence trade], *Voprosy ekonomiki,* 1975, No. 3, p. 113.
4. Hanson, *Trade,* p. 113.
5. *Novyi etap ekonomicheskogo sotrudnichestva SSSR s razvitymi kapitalisticheskimi stranami* [A new era of economic co-operation of the USSR with developed capitalist countries] (Moscow: Nauka, 1978), p. 196.
6. *Ibid.*, p. 197.

7. *Ibid.*
8. Boichenko, "Nauchno-tekhnicheskie sviazi", p. 93.
9. Irina Savyolova, "The CMEA Member Countries in the World Trade in Technology", *Foreign Trade* (Moscow), 1980, No. 1, p. 32.
10. Other examples are given in Vladislav Malkevich, "USSR Industrial Co-operation with Western Countries", *Foreign Trade* (Moscow), 1983, No. 9, p. 3.
11. Hanson, *Trade,* pp. 101, 115, and 132, citing E.I. Artem'ev and L.G. Kravets, "Izobreteniia i uroven' tekhniki" [Inventions and the level of technique], *Ekonomika i organizatsiia promyshlennogo proizvodstva,* 1979, No. 1, pp. 58-59.
12. *Novyi etap,* p. 196.
13. V. Malkevich, *East-West Economic Co-operation and Technological Exchange* (Moscow: USSR Academy of Sciences, 1981), p. 84.
14. *Novyi etap,* p. 198.
15. Hanson, *Trade,* pp. 132-133.
16. Holliday, *Technology Transfer,* p. 92, citing M. Papevich, "Regulirovanie pokupok litsenzii i 'nou-khau'" [Regulation of purchases of licences and "know-how"], *Vneshniaia torgovlia,* 1975, No. 10, p. 49.
17. N.N. Smeliakov, "I spros zavisit ot predlozheniia" [And demand depends on supply], *Trud,* 24th July 1981, p. 2.
18. In the case of some machinery and equipment, though not intermediate products, the "embodied" technology may subsequently be "disembodied" by "reverse engineering", as discussed below.
19. Antony C. Sutton, *Western Technology and Soviet Economic Development, 1917 to 1930* (Stanford, Calif.: Hoover Institution on War, Revolution, and Peace, 1968); *Western Technology and Soviet Economic Development, 1930 to 1945* (Stanford, Calif.: Hoover Institution Press, 1971); and *Western Technology and Soviet Economic Development, 1945 to 1965* (Standford, Calif.: Hoover Institution Press, 1973). For more recent examples in coalmining machinery and computers, see, respectively, US Congress, Office of Technology Assessment, *Technology,* p. 180; and Seymour E. Goodman, "Soviet Computing and Technology Transfer: An Overview", *World Politics,* Vol. 31, No. 4 (July 1979), p. 549. Only in 1965 did the USSR accede to the Paris Convention for the Protection of Industrial Property.
20. Brada, "Technology Transfer", pp. 246 and 251. East European officials report that when efforts to copy a Western product are successful it takes 5-7 years on the average from the initiation of these efforts until the copy is introduced into large-scale production, according to Business International S.A., *Selling Technology and Know-How to Eastern Europe: Practices and Problems* (Geneva, 1978), p. 6, cited in US Congress, Office of Technology Assessment, *Technology,* p. 223.
21. For further discussion of these issues, see Zaleski and Wienert, *Technology Transfer,* pp. 67-74; US Congress, Office of Technology Assessment, *Technology,* pp. 100-102; and Zdenek Drabek and John Slater, *Methodology of Data Compilation on Flows of Embodied Technology in East-West Trade,* working document, OECD, Paris, 1981, pp. 1-7.
22. For an explanation of the data base, assumptions, and methods for these OECD statistics, and comparisons with Soviet official statistics, see Drabek and Slater, *Methodology,* pp. 7-15.
23. Grant, *Soviet Machine Tools,* pp. 573-576.
24. See Philip Hanson, *Soviet Strategies and Policy Implementation in the Import of Chemical Technology from the West, 1958-1978,* (Discussion Paper No. 92; Santa Monica, Calif.: California Seminar on International Security and Foreign Policy, 1981), pp. 61-95, for a detailed tabulation of Soviet orders for chemical plant and machinery from the West during 1958-1978, by year, product group, type of machinery or equipment, contract value, capacity, supplier firm, and location in the USSR.

25. Malkevich, *East-West Economic Co-operation,* p. 125.

26. See, for example, Georges Sokoloff and Laurence Hess, "Les achats soviétiques d'usines clefs en main à l'occident", *Le Courrier des Pays de l'Est,* No. 257 (décembre 1981), pp. 32-52, and the tabulation of chemical plant and equipment contracts by Hanson cited in fn. 24 above.

27. A Soviet classification of ICAs distinguishes 1) "vertical" forms involving technology transfer through licences or delivery of plant and equipment; 2) "horizontal" forms entailing joint production; and 3) "complex" forms embracing many or all stages from research and development through investment and production to marketing. See Irina Savyolova, "East-West Industrial Co-operation", *Foreign Trade* (Moscow), 1977, No. 4, p. 23. On legal aspects, see *Pravovye formy nauchno-tekhnicheskogo i promyshlenno-ekonomicheskogo sotrudnichestva SSSR s kapitalisticheskimi stranami* [Legal forms of scientific-technical and industrial-economic co-operation of the USSR with capitalist countries] (Moscow: Nauka, 1980), pp. 119-129.

28. Malkevich, *East-West Economic Co-operation,* p. 85, and E. Iakovleva, "Production Co-operation: Properties and Potential", *Soviet and East European Foreign Trade,* Vol. XVI, No. 1 (Spring 1980), pp. 66-67, translated from *Mirovaia ekonomika i mezhdunarodnye otnosheniia,* 1979, No. 3.

29. Hanson, *Trade,* p. 111, and Lawrence H. Theriot, "US Governmental and Private Co-operation with the Soviet Union in the Fields of Science and Technology", in US Congress, Joint Economic Committee, *Soviet Economy in a New Perspective,* 94th Congress, 2d Session (Washington, D.C.: US Government Printing Office, 1976), pp. 745-766.

30. United Nations, Economic Commission for Europe, *The Role of Small and Medium-Sized Enterprises in East-West Trade and Economic Co-operation (Experience of the USSR)* (TRADE/R.421; Geneva, 14th September 1981), p. 10.

31. Holliday, *Technology Transfer,* pp. 92-93, and US Congress, Office of Technology Assessment, *Technology,* p. 103.

32. Malkevich, *East-West Economic Co-operation,* p. 129.

33. Maureen C. Smith, "Industrial Co-operation Agreements: Soviet Experience and Practice", in US Congress, Joint Economic Committee, *Soviet Economy in a New Perspective,* p. 774.

34. See A. Fedoseev, *Zapadnia: Chelovek i sotsializm* [The trap: Man and Socialism], second edition (Frankfurt: Poseev Verlag, 1979), pp. 115-117, for an emigré scientist's account of the interest of Soviet research institutes in learning through these means about Western technological developments.

35. Bruce Parrott, *Information Transfer in Soviet Science and Engineering: A Study of Documentary Channels* (R-2667-ARPA; Santa Monica, Calif.: Rand Corporation, 1981), p. 10.

36. *Ibid.,* pp. 25-27.

37. John A. Martens, "Soviet Patents and Inventors' Certificates", in US Congress, Joint Economic Committee, *Soviet Economy in the 1980s,* Part I, pp. 539-540.

38. Malkevich, *East-West Economic Co-operation,* pp. 81-82 and 93.

39. *Ibid.,* p. 82.

40. US Central Intelligence Agency, *Soviet Acquisition of Western Technology* (Washington, D.C., April 1982), p. 4.

41. Francis W. Rushing and Catherine P. Ailes, "An Assessment of the USSR-US Scientific and Technical Exchange Programs", in US Congress, Joint Economic Committee, *Soviet Economy in a Time of Change,* Vol. 2, p. 619.

42. US Central Intelligence Agency, *Soviet Acquisition,* pp. 3-5.

Chapter 4

THE IMPACT OF THE TRANSFER OF WESTERN TECHNOLOGY ON THE SOVIET ECONOMY

An assessment of the impact of the acquisition of Western technology on the Soviet economy is hampered by conceptual and statistical problems. Attempts to assess this impact by macroeconometric approaches are controversial because of deficiencies in the available data and because of simplifying assumptions in the theoretical models. The consensus of specialists is that imports of Western machinery and equipment have made only a rather modest contribution to the overall growth of Soviet industry or the Soviet economy as a whole. However, case studies show that the impact of the acquisition of Western technology has been significant for particular branches of industry (for example, chemicals), products (for instance, chemical fertilizers), and projects (such as the Volga auto plant). But the impact of imports of Western technology depends upon the speed and extent of first its absorption and then its diffusion in the Soviet economy. The scanty evidence available indicates that the USSR has considerable difficulty both in exploiting Western technology in the first facility for which it is obtained and in replicating the technology in other plants.

In acquiring Western technology by the various modes discussed in Chapter 3, the Soviet authorities presumably expect this technology will increase Soviet production, and possibly expand exports or reduce imports, to a greater extent and/or faster than these aims could be achieved with Soviet domestic technology or technology from East European countries. However, it is difficult to assess the impact of imported Western technology on the Soviet economy, for a number of reasons.

First, as explained in Chapter 3, appropriate specific quantitative data are available for commodity imports but not for modes of technology transfer like licences and industrial co-operation agreements.

Second, it is difficult to trace the effects even of imports of Western machinery and equipment. The (Western and Soviet) economic literature offers various theoretical models linking output changes to changes in the capital stock of buildings and machinery and equipment ("embodied technology"). However, the application of such models to the Soviet economy is hampered by the paucity of relevant published statistics and by the incomparability of the available data. For example, the most precise data on Soviet imports of machinery and equipment from the West (presented in Tables 10-16) are in US dollars at world market prices. Official Soviet statistics on foreign trade with the West (see Chapter 5) are expressed in "foreign-trade" *(valuta)* rubles by applying the official Soviet foreign exchange rates to transactions made in foreign currencies at world market prices. Soviet data on investment, capital stock, and output are in Soviet domestic rubles at Soviet internal wholesale prices. But (as noted in Chapter 2, Section B) these prices do not reflect relative scarcities in the Soviet economy, and the Soviet official exchange rates do not correspond to the relative purchasing powers of the ruble and foreign currencies like the US dollar. Also, most Soviet statistics on investment and capital stock are in constant (e.g. 1969, 1973, or 1976) rather than current ruble prices. Thus, many estimates and assumptions – often crude and sometimes rather arbitrary – are made in comparisons of Western and Soviet data on Soviet foreign trade with data on Soviet domestic economic activity[1].

Third, *a priori* one might expect the effect of imported Western technology on the growth of Soviet national product or total industrial production to be slight, in view of the size of the Soviet economy and the relatively small share of imported Western machinery and equipment in annual total Soviet investment in machinery and equipment. For instance, in the mid-1970s this share was perhaps 5-6 per cent[2]. However, the impact of imported Western technology cannot be measured simply by such an (order-of-magnitude) figure for the economy as a whole. For particular branches or sub-branches of industry (e.g. chemicals) the corresponding ratio may be significantly higher, and other types of commodity imports (e.g. intermediate products like pipe) and other modes of technology transfer may be important.

Fourth, there are many possible forms of impact of Western technology on the Soviet economy[3]. There is a direct net impact from the increase in output due to the use of the Western technology, instead of an alternative domestic technology, in the plant(s) in which the imported machinery or process is first employed. In addition, there may be several kinds of indirect impact:

a) "Downstream" effects will occur if the plant using imported technology produces technologically superior machinery, components, or materials (e.g. complex fertilizers) that raise the productivity of the customer enterprises using them.

b) "Upstream" effects may also occur if supplier enterprises must improve the quality of materials in order to satisfy the requirements of plants with foreign technology (e.g. better sheet steel for the Volga auto plant).

c) Another type of indirect impact is the diffusion of foreign technology when the machine or process is successfully replicated in another Soviet plant.

d) Finally, if some of the additional output attributable to the technology import is allocated to investment, this will expand capacity and output even if that new capacity does not include any Western technology.

Fifth, some Western writers have suggested that, along with these positive effects, imports of Western technology may also impose resource costs on the Soviet economy because such imports require complementary inputs – such as skilled labour, high-quality materials, and research and development resources – that must be diverted from other activities[4]. But in a "full employment", "tautly planned" economy, any project, with or without Western technology, competes for scarce resources. Because projects with foreign technology command special attention from planning and administrative agencies, it is reasonable to expect that these agencies carefully consider the corresponding domestic resource requirements, and conclude they are justified, before undertaking the technology imports. Of course, it is possible that – as in the case of entirely domestic projects – planners' calculations may prove inaccurate and overoptimistic, resulting in shortages that delay construction and/or reduce output. If so, the likely high priority of projects with Western technology may lead to the reallocation to them of resources originally committed elsewhere in the economy. But there is little published evidence of such serious planning errors and subsequent substantial adjustments[5].

A related argument is that reliance on Western technology in a particular industry weakens Soviet R & D capabilities in that field, or at least shifts them from work on genuine innovation toward merely adapting foreign technology. However, the scale of Western technology imports has not been large enough to displace Soviet domestic efforts at research, development, and innovation. Also, "adaptive" R & D may in turn stimulate "creative" R & D[6].

Thus, a truly satisfactory appraisal of the impact of the acquisition of Western technology on the Soviet economy is not possible because of the many dimensions of impact, the limited data available, and the measurement problems involved.

The remainder of this chapter considers three aspects of this impact that are covered, to some extent, in published studies. Attempts to estimate the impact of technology transfer by macroeconomic approaches are examined in Section A. Evidence from selected branch case studies is discussed in Section B. Information on Soviet problems in the assimilation of Western technology is appraised in Section C.

The impact of the transfer of Western technology on Soviet foreign trade is the subject of Chapter 5.

A. MACROECONOMIC APPROACHES

Macroeconomic estimates of the impact of Western technology on the Soviet economy are controversial because they rely on simplifying assumptions of disputed validity as well as on incomplete and imperfect statistical data. Hence, these estimates are the subject of criticism and disagreement among experts on technical grounds. The dominant view among specialists now is that the aggregate impact of the transfer of Western technology on Soviet industry cannot really be measured satisfactorily but is likely to be rather modest (and the impact on the Soviet economy as a whole, of which industry is only a part, even more so). The aims, methods, results, and shortcomings of such estimates have been examined at length in earlier studies, including one exclusively on this subject prepared especially for the OECD[7]. Hence, only a few of the most important estimates are summarised briefly here.

If Western machinery and equipment are imported on credit, this could increase the volume of capital formation in the Soviet economy, to an extent dependent on the amount of capital goods imports so financed and their productivity[8]. However, rough estimates suggest that on an annual basis during 1970-1977 imports of Western machinery and equipment were equivalent to perhaps 4-5 per cent of total Soviet investment in machinery and equipment[9] (and a smaller share of total fixed investment, which also includes investment in buildings). Moreover, not all of these imports of Western machinery and equipment were obtained on credit.

This kind of impact of technology transfer was examined in a study by Desai. She estimated Solow-Swan and Harrod-Domar models of Soviet economic growth using a constant elasticity-of-substitution (CES) production function. In a test in which purchases of Western equipment were increased, she found that the internal rate of return for such incremental investment varied from 0.8 to 4.41 per cent, depending on the time span and the amount borrowed – a yield much below the interest rates on Western credits for Soviet purchases of machinery and equipment. This result supports the view that, by itself, the net foreign resource inflow from credit-financed Western machinery and equipment imports has a minor effect on the Soviet economy[10].

However, Desai's estimates have been criticised as too low, due to an unrealistic assumption of a constant rate of technological change during 1950-1975 and, as a result, an unduly low elasticity of capital-labour substitution and an excessive decline in the estimated marginal productivity of capital.[11]

Moreover, Desai's model assumes that the Western machinery and equipment imported with and without credit do not differ technologically (or in any other sense) from Soviet machinery and equipment. In contrast, studies by Green and Levine started from the hypothesis that imported Western machinery and equipment are technologically more advanced, and of greater marginal productivity in combination with other inputs, than the closest available Soviet machinery and equipment. Therefore, Green and Levine attempted to measure the differences in the marginal productivity of Western versus Soviet machinery and equipment in Soviet industry as a whole and in particular branches of it, by estimating Cobb-Douglas production functions with Soviet and Western machinery and equipment as separate inputs. For 1960-1974 they found that the marginal productivity of Western machinery and equipment was much greater than that of Soviet machinery and equipment. For example, in one set of results, the former was 14 times as large as the latter for industry as a whole; 8 times as large for chemicals and petrochemicals, and for petroleum products; and 22 times as large for machine building and metal working. These results were used in an econometric model of the Soviet Union to estimate how Soviet industrial growth during 1968-1973 would have differed if there had been no increase in imports of Western machinery and equipment. Green and Levine calculated that over the entire period of five years the growth of Soviet industrial output would have been 2.5 percentage points lower (29.6 per cent versus 32.1 per cent)[12].

The work of Green and Levine has been criticised by several specialists on various counts as overestimating the difference in the marginal productivity of Western versus Soviet machinery and equipment and thus the impact of imports of Western machinery and equipment on the growth of Soviet industrial production.

For example, there are shortcomings in the basic data series used. In the industrial output series, new products are neglected. In the labour input series, employment rather than the number of worker-hours is used. In the capital stock series, questionable rates are used for the conversion of import data in current foreign-trade rubles into some equivalent of the constant domestic wholesale prices in which the Soviet capital stock is valued. Imports from Eastern Europe are included in some of the figures for "Western" machinery and equipment. The assumed lag between delivery of Western machinery and equipment and the start of production with them is much too short.

In the specification of their model Green and Levine treat Western machinery and equipment as a separate and essential input, like labour or domestic machinery and equipment – in effect assuming that Soviet industrial production would be impossible without Western machinery and equipment. Also, there is no separate term for technological progress. Hence its effect on economic growth is attributed to the growth of physical inputs, raising the estimated contribution of imported machinery and equipment to Soviet industrial growth. The conclusions of Green and Levine are based on a data set for a relatively small number of years, and their results for industry as a whole are aggregated from a small number of branches. Finally, the apparent superior marginal productivity of Western machinery and equipment may be due in part to a higher priority in the allocation of scarce complementary inputs (labour, materials) to projects with Western equipment[13].

Still another macroeconomic approach is used by Gomulka. His Kalecki-type growth model aims to estimate the effect of imports of Western machinery and equipment in raising labour productivity, and thus output, in Soviet industry. In his model the effect depends on:

1. The difference in the productivity of Western versus Soviet machinery and equipment;
2. The share of the superior imported machinery and equipment in total investment; and
3. The speed of installation and the extent of utilisation of Western versus domestic machinery and equipment.

By "rule-of-thumb" calculations, Gomulka estimated that a 1 per cent increase in the share of Western machinery and equipment in total Soviet investment in machinery and equipment causes an increase in the growth rates of Soviet industrial labour productivity and output by about 0.1 per cent. Thus, if imports of Western machinery and equipment account for, say, 4 per cent of annual Soviet investment in machinery and equipment, these imports will add 0.4 per cent to annual Soviet industrial growth[14].

However, Gomulka's approach entails the assumption that once Western machinery and equipment are put to use in Soviet industry, the associated labour productivity in the USSR is the same as in the West. But the associated output and labour productivity (output per worker or per worker-hour) in the USSR may often be lower, because of assimilation problems (discussed in Section C below) and/or shortages of materials. To the extent that this is so, Gomulka-type estimates will overstate the likely impact of imports of Western machinery and equipment on Soviet industrial growth.

These and other attempts at macroeconomic estimates of the impact of imports of Western machinery and equipment on the growth of the Soviet economy, or Soviet industry or branches of it, do not provide reliable conclusions, because of deficiencies in the data used and because of simplified and often doubtful assumptions (and corresponding econometric specifications). These exercises have been more useful in identifying questions and exploring them than in providing definitive answers to them. Western machinery and equipment are a minor element in total Soviet investment and capital stock. It is reasonable to expect that because of their technological superiority Western machinery and equipment could have a disproportionately large effect in expanding Soviet industrial output. But it is not possible to measure this effect satisfactorily with the available statistical data and the current state of the art in econometrics. The most that expert opinion can offer on the subject is a tentative, speculative, order-of-magnitude conjecture that imports of Western machinery and equipment did not contribute more (and may have contributed less) than half a percentage point of the annual growth of Soviet net industrial output in the 1970s[15].

B. CASE STUDIES

Although the impact of technology transfer from the West on the overall growth of Soviet industry or the Soviet economy is thought to be relatively small, this impact might be more significant for particular branches of industry (e.g. chemicals), products (e.g. chemical fertilizers), or projects (e.g. the Volga auto plant). Case studies may provide information on the nature and size of the effects of technology transfer from the West, on how these effects were achieved, and on the factors limiting them. However, the available case studies differ considerably in their aims, scope, depth, data base, methodology, time period covered, and form of publication. Hence, they do not furnish comparable assessments of the impact of the transfer of Western technology on different branches and sub-branches of the Soviet economy[16]. Nevertheless, despite their individual weaknesses and their non-comparability, such case studies are a necessary complement to the macroeconomic approaches considered in Section A.

This section summarises some of the quantitative and qualitative evidence from selected studies on the transfer of Western technology to the Soviet chemical, motor vehicle, machine tool, energy, and forest products industries.

1. Chemicals

The dominant mode for the transfer of Western technology to the Soviet chemical industry has been large-scale purchases of machinery and equipment, often for turnkey projects (some at existing chemical complexes and some at new sites) and frequently complemented by industrial co-operation agreements. Imports of Western plant and equipment for the Soviet chemical industry totalled some $5 billion in 1961-1975. Of the total, about $3.6 billion, or approximately three-fourths, occurred during 1971-1975 and was equivalent to about one-fourth of total Soviet investment in chemical plant and equipment during the period[17]. According to official Soviet foreign trade statistics, the share of chemical equipment in total Soviet imports of machinery and equipment from the West reached over 36 per cent in 1977[18].

The heavy reliance on Western chemical technology and equipment is due partly to Soviet weaknesses in the research and development stages for many chemical products and processes, partly to problems in carrying promising development through to large-scale production, and partly to the inability of Soviet metallurgy and metalworking to furnish adequate tanks and tubes for the chemical industry. Soviet efforts in the 1950s and 1960s to develop improved processes or equipment for such important chemical products as ammonia, acrylonitrile, caprolactan, and polyethylene resulted in relatively high-cost, low-volume installations obsolete by world standards by the time they were commissioned. Also, the quality of Soviet output of such products as dyes, synthetic fibres, plastics, and industrial rubber goods has often been criticised in the Soviet technical press[19].

Despite a growing ability to supply some types of chemical equipment and related technology, East European countries cannot fill the gap between Soviet production and needs. The chief East European contributions to meeting Soviet requirements have been in basic chemicals and fertilizers, for which the production technology is less sophisticated than the technology for complex fertilizers, petrochemicals, and synthetic materials supplied by the West. However, some chemical technology furnished to the USSR by Eastern Europe, for example for sulphuric acid and urea, is of Western origin, having been provided earlier by Western firms to Poland and Czechoslovakia, respectively[20].

The bulk of Western chemical equipment bought by the USSR has been for the production of fertilizers (ammonia and urea) and artificial fibres (polyester fibre, polypropylene, cellulose triacetate, and intermediates for making nylon and other synthetic fibres). Equipment for fertilizers and artificial fibres accounted, respectively, for about 40 and about 20 per cent of the total value of Soviet chemical equipment purchases from the West in 1971-1975, for example. Other important categories are equipment for production of plastics (polyethylene, polyvinyl chloride, and polystyrene) and of synthetic rubber (linked to the expansion of output of tyres and other rubber components of motor vehicles)[21].

Western equipment has played an important role in the expansion of the Soviet chemical industry's capacity and output. For instance, 72 per cent of the increase in ammonia production capacity in 1971-1975 was due to Western equipment. In 1975 plants from Western countries produced the following shares of total Soviet output: acrylic fibres, two-thirds; polyethylene, three-fifths; and polyvinyl chloride, one-third[22]. Projects under compensation agreements with Western firms accounted for 55 per cent of the increase in production capacity for ammonia during 1976-1981. The corresponding figure for carbamide was 30 per cent[23].

The import of Western equipment, including whole installations for turnkey projects, has shortened construction periods, reduced investment costs, and cut production costs, in comparison with facilities using Soviet equipment instead[24].

The period from placement of Soviet orders to initial operation of Western-supplied plants commissioned in 1971-1975 was generally 3-5 years, while the average duration of construction for all Soviet chemical plants is about 8 years. For instance, a Western-equipped ammonia plant at Novgorod commissioned in 1975 was built in a little less than 3 years, whereas earlier construction of a Soviet ammonia unit at the same facility took 8 years, even though its production capacity was less than half that of the imported installation.

Western process technology has facilitated Soviet use of more economical feedstocks and processes. For example, an imported facility using urea, instead of calcium cyanamide, was expected to save 50 per cent of the cost of producing melamine for dishes, surface coatings, and adhesives. The one-stage process for butadiene (a major intermediate used in rubber production) in a plant imported from the West is likely to use at least 20 per cent less butane per ton of butadiene than the two-stage process in Soviet-built plants.

Western technology for ammonia production allows a saving of 95 per cent in electric power consumption per ton of ammonia, compared to the technology in earlier Soviet-designed units.

Western ammonia plants require only 10-30 per cent as many workers per ton of ammonia as do older Soviet plants, and substantial labour input savings are expected from imported Western plants for other types of chemicals.

However, there are several problems in the construction and operation of Western-supplied plant and equipment that reduce their impact on the Soviet chemical industry. Soviet organisations lack experience in the installation, operation, and maintenance of complex chemical equipment imported from the West. For instance, completion of facilities with Western equipment is delayed by poor-quality welding, errors in installation of equipment and insulation, and improper operation and servicing of equipment and instruments. Once in operation, these facilities often have difficulty obtaining materials, spare parts, and qualified repair personnel[25].

Most case studies, of other industries as well as of the chemical industry, aim (at most) to trace only direct impacts of Western technology on production capacity, output, and costs for the particular product or product group. In contrast, in a study of the import of Western technology for mineral fertilizer production, Hanson tried to estimate also the indirect "downstream" effect from the use of this fertilizer in agriculture[26].

The Soviet aim was to increase output of complex fertilizers (which include mixtures of at least two of the three elements nitrogen, phosphorus, and potassium, or their compounds, in concentrated form). Soviet policy stressed imports of Western plants and know-how for the development of nitrogeneous (especially ammonia) and compound fertilizer production, and relied largely on domestic plants for phosphate and potash fertilizers. The fertilizer production technology imported from the West was quite advanced and far ahead of Soviet research and development in that field.

Hanson roughly estimated the share of total mineral fertilizer supplies produced in plants with Western equipment in the mid-1970s at one-fourth. Combining this production with different Soviet estimates of the contribution of fertilizer to net agricultural output, he estimated very roughly that fertilizer from these plants may have been responsible for between 0.4 and 2.2 per cent of Soviet net agricultural output in the mid-1970s. However, since the construction of these plants also involved inputs of Soviet equipment, labour, and materials, the contribution of Western technology as such was less than these figures would indicate.

Although imaginative and carefully explained, Hanson's results are, as he states, only "speculative and approximate" because of the paucity of relevant data (and thus the many assumptions to be made) about both the share of Western technology in Soviet fertilizer production and the contribution of fertilizer to Soviet agricultural output.

2. Motor Vehicles

Western technology also played an important role in the development during the 1960s and 1970s of Soviet production of both passenger cars and trucks.

By the mid-1960s the Soviet Government recognised that output of passenger cars (only 201 000 in 1965) was extremely inadequate. Press accounts even reported the frequent use of trucks for business transportation of managerial personnel. Also, the regime wanted to offer the possibility of individual passenger car ownership to a considerable segment of the elite, thereby providing an outlet for some of the savings they were accumulating as a result of widespread shortages. The Soviet authorities decided on a six-fold increase in the output of passenger cars, to 1.2 million units, by 1975. But such an expansion could not be attained by gradual increases in the output of existing Soviet auto plants, which were in any case producing vehicles technologically obsolete (in engineering, performance, and quality) by Western standards. The Soviet Government thus sought Western designs, machine tools, and production know-how for two passenger car programmes. One was the tripling of the production of Moskvich cars – through expansion of the Moskvich plant in Moscow and the conversion to Moskvich production of the Izhevsk Machinebuilding Plant – with the assistance of the French firm Renault. The other was the construction of a large new facility with an annual output of 660 000 cars, the Volga Automotive Factory (VAZ) at Tol'yatti, with the assistance of the Italian firm FIAT[27].

Despite the great reliance on Western assistance, neither the Moskvich nor VAZ programmes involved turnkey projects. In the Moskvich programme, the product is mostly a Soviet design, though the engine is a close replica of a West German BMW engine. The VAZ vehicle is based on a FIAT model, but it was substantially redesigned, with full Soviet participation, for Soviet conditions, including poorer roads and colder climates than in Western Europe. Soviet organisations also designed all the buildings and supplied a significant amount of the production equipment for both the Moskvich and the VAZ projects[28].

A scientific and technical co-operation agreement between the USSR and FIAT in 1965 was an important first step toward the latter's final contract for the VAZ project. Under this agreement, the Soviet side learned about FIAT's research and development work, production techniques, finished products, and future plans, including the FIAT-124 car.

Under the contract for VAZ signed in 1966, FIAT sold the USSR manufacturing rights for a Soviet version of the FIAT-124, called the "Zhiguli" in the USSR and the "Lada" abroad. FIAT agreed to furnish designs for the production process; to specify what Western machinery and equipment should be used; to supervise assembly and installation of all imported equipment and to guarantee its successful operation; and to train Soviet personnel in the production process. The USSR obtained credit of about $322 million from the Istituto Mobiliare Italiano to cover 90 per cent of the cost of imported machinery and equipment. A large part of the Western machinery installed in VAZ was made by other Western firms as subcontractors for FIAT, which also acquired Western licences for VAZ. Soviet-manufactured materials were sent to FIAT's Turin factory for quality control tests. About 2 500 Western specialists, including 1 500 from FIAT itself, went to Tol'yatti, and about 2 500 Soviet technicians were sent to Italy[29].

FIAT's help was thus critical for the success of VAZ. Without it, the USSR would have had great difficulty choosing among Western products and processes and combining them into an efficient auto-producing facility. The USSR has sought to maintain its ties with FIAT by successive renewals of their scientific and technical co-operation agreement, and by FIAT involvement in Soviet tractor production.

The FIAT-assisted VAZ accounted for 57 per cent of total Soviet passenger car production in 1978. In addition, part of the output of Moskvich cars was due to Renault's role in modernising and expanding plants producing them[30].

In late 1983, Renault signed a new contract to assist in the design and testing of a new model of the Moskvich, with front-wheel drive, to go into production in 1986 at the Moskvich Plant in Moscow. Renault will help Moskvich develop the prototype and will provide technical and design expertise in the machining of mechanical parts, body-stamping, assembly, and painting. Renault will receive $36.8 million for its assistance, and related sales by other French firms are estimated at $122 million[31].

Large amounts of Western machinery and equipment were used also in the Kama Automotive Factory (KamAZ) at Naberezhnye Chelny (subsequently renamed Brezhnev) on the Kama River. This plant was designed for annual production of 150 000 diesel trucks (and engines) of 8-ton capacity and an additional 100 000 diesel engines for installation in trucks and buses produced at other plants. The USSR sought to obtain from the West the bulk of the equipment and the services of a major Western firm or consortium to perform the co-ordinating role FIAT played for the VAZ plant, but it was unsuccessful. Foreign procurement was limited to providing equipment with a convertible currency cost of about $1 billion for the first stage of the Kama plant's development, that is, production of 75 000 trucks and 40 000 extra engines by 1975. The Soviet machine tool industry supplied most of the equipment for the completion of the second stage by 1980. The USSR announced output of 41 000 KamAZ trucks (about 5 per cent of total truck production) in 1978, and the plant was scheduled to reach designed capacity of 150 000 in 1983[32]. However, KamAZ produced only 85 000 trucks in 1982, and achievement of the goal of 150 000 has been postponed to 1985 at the earliest and possibly later[33].

The Kama plant is much more mechanised and automated than other Soviet truck plants. Partly as a consequence, construction of the first phase of KamAZ involving considerable Western machinery and equipment fell behind schedule. Work on the site began in 1968, and the start of production was scheduled for 1974 but in fact occurred in 1976. The Soviet press, as well as participating Western firms, reported insufficient co-ordination of the specifications of related machinery and equipment purchased from different Western suppliers; shortages of skilled labour and high turnover rates at the site; delays in shipments of Soviet materials and construction components; deliveries of equipment before the space was ready for its installation, requiring the construction of unanticipated temporary storage facilities; and, as a result of the preceding problems, large cost overruns.

At least some of these problems in the KamAZ project stemmed from the absence of a Western firm serving – as FIAT did in the VAZ project – as general consultant; co-ordinator of purchases of Western machinery, equipment, and licences; and source of construction and production know-how[34]. However, the VAZ project also suffered from construction delays due to shortages of Soviet labour, materials, and equipment.

Once production began at VAZ and KamAZ, shortages of satisfactory materials and components retarded output growth and reduced the quality and reliability of completed vehicles. A major reason was that the (for the Soviet Union) extremely advanced technological level of the Zhiguli required types and qualities of components, tyres, lubricants, coolants, etc., never previously made in the USSR and for which new supplier plants had to be built[35]. Similar problems occurred at KamAZ[36].

3. Machine Tools

The USSR has obtained Western machine tool technology through a variety of "passive" and "active" ("non-negotiated" and "negotiated") transfer mechanisms. These include:

1. The study of Western published literature;
2. Participation in business and intergovernmental meetings;
3. Acquisition of Western licences for design and production;
4. Purchases of Western machine tools (sometimes in the form of complete factories or processing sections); and
5. Industrial co-operation agreements.

The last three modes – licences, commodity imports, and industrial co-operation – have proven the most effective, because Western firms provide more recent technology and also production know-how in the use of the machine tools.

One way of assessing the significance of Western machine tools for the Soviet economy is to compare imports of Western machine tools with Soviet machine tool production or consumption (production plus imports minus exports). Such a comparison for 1971-1975 in units showed that Western supplies of machine tools accounted for 15-30 per cent of total Soviet machine tool imports, but only 1-2 per cent of total Soviet machine tool consumption, because of the large Soviet domestic production relative to foreign trade. For example, in 1975 Soviet domestic production was 231 000 units and imports from the West were only 2 200 units. However, these figures understate the shares of Western machine tools, because some metalcutting machine tools are classified in Soviet import statistics under "production equipment for automobile factories", which is supplied almost exclusively by the West. Furthermore, such comparisons in physical units understate the importance of Western machine tool imports to the USSR because Western machine tools are more advanced and complex than Soviet machine tools and also than East European machine tools imported by the USSR. According to Soviet statistics, the average value of machine tools imported from the West is double to triple that of machine tools imported from Eastern Europe[37].

Of the total value of Soviet imports of machine tools from the West in 1970-1977, according to Western statistics, metalcutting tools accounted for about half; metalforming tools, a little more than one-third; and transfer lines (which can combine both types), the remainder. Within the total value of metalcutting tools, conventional grinding tools and automatic lathes each represented about one-fourth, and numerically-controlled machine tools of all types about one-fifth[38].

Machine tool imports from the West made an important contribution to the VAZ passenger car plant and the KamAZ truck plant, discussed in the preceding sub-section. Without Western machine tools, these projects could have been completed only in a much longer time period, through great demands on the already severely strained Soviet domestic machine tool industry, and with a resulting quality of final products well below that of Western vehicles[39].

In addition, imports of Western machine tools met spot shortages of special-purpose machine tools in other parts of Soviet industry. Also, imports of machining centres provided advanced capabilities beyond the current Soviet state of the art. For instance, imports of Western high-precision grinding equipment made possible the production of miniature bearings with important civilian and military uses[40].

Finally, through industrial co-operation agreements with Western firms the USSR is obtaining design technology for conventional and advanced machine tools and production technology for advanced types. In the conventional category, these agreements stress automatic lathes, grinding machines (both internal grinders to make miniature bearings and high-speed grinders), and electronic measuring devices. In the advanced category of numerically-controlled machine tools, the USSR is receiving technical assistance for machining centres; milling, grinding, and boring machines; and programming[41].

4. Energy

In regard to coal, nuclear energy, and electric power, Western expert technical opinion is that technology transfer from the West has been modest, with little impact on the corresponding Soviet industries[42].

Most of the Western research on technology transfer in the energy sector has therefore focused on the oil and gas industries. This research has sought primarily to trace how Western technology improves Soviet production or transportation capabilities, without attempting quantitative estimates of how much additional Soviet production capacity, output, or exports can be ascribed to imports of Western technology.

The USSR has been very selective in the purchase of Western equipment and technology for its oil and gas industries, although part of the explanation lies in the restrictions by some Western governments on the sale of some items to the USSR[43].

Drill pipe imported from Western Europe and Japan has enabled the Soviets to drill deeper than with domestically-produced pipe. It is not possible to estimate the amount of such imports, because they are not reported separately in Western or Soviet foreign trade statistics. But it is reasonable to assume that this pipe is reserved for the deeper wells that currently account for 5-10 per cent of Soviet oil producton.

The USSR in 1978 signed a contract with the US firm Dresser Industries for a complete plant and all designs and processes to manufacture 100 000 drill bits per year, including 86 000 tungsten carbide insert bits.

The Soviet Union has imported from Canada and Finland drilling rigs that permit faster, deeper, and wider drilling than Soviet rigs.

Although the USSR has a substantial capability to produce electric submersible pumps, Soviet models have lower capacities and require more maintenace than Western models. Hence, until 1978, the USSR imported such pumps from the United States, the only other producer. In Soviet conditions, US pumps need a major overhaul after 3-6 months of use. However, US firms refused to train Soviet personnel in pump repair. Unless Soviet personnel learned how to repair the equipment without the help of the suppliers, the US pumps must be discarded when they fail, and none bought through 1978 would still be in use.

The Soviets have also purchased or contracted for offshore drilling, exploration, and production equipment from several Western countries, for use in Baltic and Arctic waters and off Sakhalin Island.

The USSR has bought complete oil refineries from the West, although the technologies were generally not advanced except for the use of minicomputers and microprocessors in control systems.

Sophisticated automation systems have also been acquired from the West for oilfields (e.g. Samotlor and Fedorovsk) and gasfields (e.g. Orenburg). The equipment used in these systems only recently went into production in the CMEA countries. Thus, purchases from the West provided capabilities far advanced over those otherwise available to the USSR. The systems obtained for the Samotlor and Fedorovsk fields were especially important, because these fields are in the critical secondary recovery stage of production.

Soviet gaslift efforts have been enhanced by a $200 million deal in 1978 with two French firms for equipment for 2 400 wells, including gas compressors, high-pressure manifolds, and control valves. However, this amount of equipment will serve only about 20 per cent of Samotlor's wells, or perhaps only 1-3 per cent of current Soviet gas production.

The USSR has imported West European and Japanese pipe of 40-inch and greater (often 56-inch) diameter for gas pipelines[44]. The Soviet steel industry produces 40-, 48-, and 56-inch pipe, but large imports reflect both inadequate Soviet production capacity and the lower quality of Soviet pipe in regard to yield strength, wall thickness, and general workmanship (especially welding). Soviet pipe is apparently inadequate for lines operating at more than 55 atmospheres, and new Soviet lines operating at 75 atmospheres require imported pipe[45].

Large-diameter pipe is not identified separately in Western (SITC) foreign trade statistics, and it has not been shown separately in Soviet foreign trade statistics since 1975. For 1971-1975, Campbell estimated that imports of large-diameter pipe for oil and gas pipelines amounted to about 6.3 million tons, equivalent to about two-thirds of the total amount of pipe installed in large-diameter lines. These data refer to both oil and gas pipelines, but it is likely that the reliance on imported pipe was even greater for gas lines alone, since the quality demands for pipe used in gas lines are higher than for pipe used in oil lines.

In addition to buying pipe abroad, the USSR has purchased a seamless pipe manufacturing plant from the West, with an annual capacity of 170 000 metric tons (equivalent to about 7 per cent of 1976 Soviet output of 40-inch and larger pipe).

Soviet industry has also had difficulty developing and producing compressor equipment for gas pipelines. Soviet efforts to make turbine-powered compressor units have been characterised by errors, long delays in mastering the production of large-size units, and frequent shifts in development tactics in response to these problems. The average size of Soviet compressor turbines is still relatively small in comparison with Western equipment – 4, 6, and 10 megawatts versus 16 and 25 megawatts. Also, the Soviet units require complicated and costly installation at the site. Plans for compressor output have often been seriously underfulfilled. Hence, the USSR has imported gas-turbine-powered compressors from the West for the Soiuz Pipeline from Orenburg to Eastern Europe and for other lines[46].

For the gas pipeline network from the Urengoi fields in Northern Siberia, which includes five domestic lines and one line to Western Europe, the USSR is importing from the West about 10 million tons of 56-inch pipe, 125 compressor stations, and related equipment, at a total cost estimated at about $15 billion. Urengoi No. 6, the 2 800 mile line to Western Europe through Czechoslovakia, alone involves imports from the West of about 3 million tons of pipe and 41 compressor stations (each with three 25-megawatt turbines). For Urengoi No. 6, the total cost of the pipe, compressor stations, and related equipment imported from the West is estimated at about $5 billion[47].

Western pipe and compressors have made a substantial contribution to Soviet gas pipeline transport by reducing construction and installation costs and by increasing operating productivity through higher pressures and through higher utilisation rates due to fewer and shorter interruptions of service from pipe leaks and equipment breakdowns. However, the gains from the use of Western pipe and equipment in the Soviet gas industry have been limited by such chronic problems of the Soviet economy as poor planning and management and lack of manpower, fuel and power, transport, and infrastructure like repair and storage facilities, housing, and roads[48]. Soviet gas pipelines built with Western pipe and equipment transport gas not only for domestic use but also to Eastern Europe and Western Europe. Soviet oil production may peak, while gas production rises sharply, and gas exports to Western Europe, at likely quantities and prices, will become a significant earner of convertible currency for the USSR[49]. (See Chapter 5 for further discussion of gas exports.)

5. Forest Products

In the case of forest products, the Soviet Union has sought Western technology for several reasons[50]. One is to equip large-scale enterprises with labour-saving machinery. A second is to achieve more efficient processing of wood into wood products, including utilisation of chips and waste. A third is to operate in adverse environmental conditions in Siberia. The fourth is to raise convertible currency earnings from exports, by changing the product-mix to increase sales of higher-value products (e.g. composition board) and by improving the quality of pulp and paper goods.

For these objectives, the Soviet Union has imported from the West machinery and equipment for the paper, pulp, and wood-processing industry (amounting to about $170 million in 1979 and $97 million in 1982, according to Annex Table A-8). Turnkey projects and other forms of industrial co-operation, as well as compensation agreements, have been important methods of technology transfer. For example, under a compensation agreement, British, French, Swedish, and Finnish firms participated in the development of the Ust-Ilimsk timber complex, supplying complete installations for saw mill and pulp-making operations, technical documentation, supervision of construction and assembly, and training of personnel[51]. The USSR also has negotiated several compensation agreements with Japan, providing for Soviet purchases of Japanese machinery and equipment for transportation, road-building, timber-cutting and processing, and paper and paperboard manufacturing.

Braden attempted to measure the superior marginal productivity of foreign over domestic machinery and equipment in the Soviet forest products industry, using the methodology of Green and Levine (discussed above in Section A of this Chapter). She found that the marginal productivity of Western machinery and equipment was somewhat higher, but that the difference was much smaller than the differences which Green and Levine calculated for some other branches of industry. However, Braden acknowledged the limited validity of her results, due to lack of appropriate data, simplified assumptions (e.g. no time lag between the import of machinery and its use in production), and evidence of autocorrelation and multicollinearity in her regression results.

From a study of particular projects and products, Braden concluded that Western (embodied and disembodied) technology had a significant effect on Soviet production (e.g. of small-diameter logs), utilisation of by-products and thinnings, improvement of the quality of exports, and application of environmental protection (e.g. anti-pollution) measures. However, she also found that, for instance in the Bratsk forest industry combine, the benefits of imported machinery were reduced by labour shortages, poor co-ordination of machinery deliveries and building construction, and, as a result, start-up delays.

In summary, case studies of the transfer of Western technology in the five fields discussed in this section show significant contributions to production capacity, output, cost reduction, quality improvement, and convertible currency exports – although the relative importance of these different effects varies across the several fields. The roles of different modes of technology transfer also vary in the five cases. For example, turnkey plants and/or industrial co-operation were important for chemicals, motor vehicles, and forest products, but not for gas pipelines, for which commodity imports of pipe and compressors have been the chief method of technology transfer.

The technology transfer from the West to the USSR examined in these case studies raised the technological level of the particular branches of Soviet industry. But technological progress in these activities has continued in the developed market economies. Hence, the specialised technical and economic literature does not indicate any significant narrowing of the technological lag of the USSR behind the West in these fields.

C. SOVIET ASSIMILATION OF WESTERN TECHNOLOGY

The impact of imports of Western technology depends, *inter alia,* upon the speed and extent of its assimilation by the Soviet economy. The assimilation process has two main phases. The absorption phase involves the successful exploitation of the Western technology in the first facility for which it is acquired. The diffusion phase entails the replication of the Western technology in other plants. Weaknesses in either phase reduce the impact from the acquisition of Western technology and can thus influence Soviet decisions about the scale, modes, and fields of subsequent technology imports from the West.

1. Absorption

Our knowledge of Soviet experience in the absorption of Western technology is relatively limited.

No comprehensive or systematic evaluations of Soviet absorption experience have been published in the Soviet general or specialised press. However, Soviet publications reveal various instances in which first the completion of facilities with Western machinery and equipment and then the operation of these facilities were hampered by poor planning and co-ordination, reflected in shortages of labour, materials and components, fuel and power, and transport. A number of examples were reported in the case studies reviewed in the preceding section[52]. Also for these reasons some Western licences have not been "mastered" by Soviet industry[53]. The underlying causes – poor planning and shortages of complementary inputs – are common in the Soviet economy. But, according to a Soviet study of the investment process, these problems hamper the absorption of new technology to a significantly greater extent when it is of foreign, rather than domestic, origin[54].

Some information about Soviet experience in the absorption of Western technology can be obtained from Western firms that provide machinery and equipment, including turnkey plants, and technical assistance to the USSR. Well-conducted surveys of such firms, with good response rates, can provide considerable information about the parts of the technology transfer process in which these firms are directly involved, for instance initial contacts; contract negotiations; and delivery, installation, and commissioning of equipment. But these Western firms are likely to know little about the subsequent Soviet use of the technology in the facility (e.g. extent of capacity utilisation, labour requirements, output levels, and quality achieved) – unless the firms have continuing close contact with the facility under an industrial co-operation agreement.

Also, Western specialists may be able to compare Soviet and Western performance in the use of Western machinery and equipment, but they seldom have the information to compare Western technology versus Soviet projects with domestic technology. However, the latter comparison is more relevant for an assessment of the superiority of Western technology and its impact on Soviet economic growth. Although Western technology may be used less effectively in the USSR than in the West, the Western technology (used less effectively) may still be superior to the alternative Soviet technology.

Nonetheless, some useful evidence on Soviet absorption experience is provided by parallel surveys of British and West German firms supplying complete plants for the Soviet chemical or machine building industries[55]. Most of the contracts covered design, engineering, delivery of equipment, supervision of installation, and on-site testing. Soviet organisations generally did the construction of production and infrastructure facilities, and the installation work. Some of the projects were for new plants and some for the expansion of existing facilities. Contracts typically involved several licensors and subcontractors on the Western side, and negotiations were therefore complex and lengthy.

The projects covered in these surveys had lead times of 6-7 years from the initial Soviet inquiry until the commissioning of the project – which the respondents judged as 3½-4 years longer than the corresponding lead times of 2¼-3½ years for comparable projects in Western Europe. Of the total difference of 3½-4 years, some 9-10 months was due to a longer negotiating period, and about 2½-3 years to a longer interval between the signing of the contract and the handing over of the project to the Soviet side. Both components of lead time should be considered in order to assess the reasons for the delay between a Soviet decision to seek Western technology and its introduction into production.

Several factors were responsible for the period between initial inquiry and contract. Initial Soviet inquiries were often vague, requiring clarification before Western firms could decide whether to start negotiations. Then the Soviet side often requested unusually detailed documentation, including evidence that the Western equipment was the most advanced in the world and commercially proven in the West. The separation on the Soviet side of the foreign trade organisation and the final user, the branch ministry, tended to make technical and commercial aspects of negotiations somewhat distinct and sometimes sequential. However, for the Soviet side the long negotiating period had technical benefits in the acquisition of a large amount of technical information from formal proposals and from discussions, and commercial benefits in the achievement of more favourable price and credit terms.

The interval between the signing of the contract and the delivery of equipment was sometimes lengthened by Soviet insistence on inspection of equipment in the exporter's country before its shipment to the USSR.

More significant was the delay from the delivery of the equipment to its installation and use in production. The main reasons were shortages of construction labour, and poor qualifications and motivation of available personnel; shortages of transportation and of complementary inputs from Soviet industry, like high-grade steel and electronic components; and lack of relevant manufacturing know-how.

Western firms usually have little, if any, knowledge of whether Soviet management maintains, exceeds, or falls short of the guaranteed-output levels achieved in the trial period before the plant is formally turned over to the Soviet side by the Western contractor. In Western countries, new plants often operate above the guaranteed level, because of subsequent modifications and rising productivity of labour and equipment. In some cases, survey respondents believed that the Soviet plants were operating below the guaranteed levels, or below the above-guaranteed levels that would have been attained in the West. The reasons cited included non-completion of related (upstream and downstream) plants, shortages of materials and transport, use of unsuitable materials, failure to keep working areas sufficiently free of dirt and dust, and lack of personnel qualified to operate highly automated plants. However, the labour force in these plants was sometimes much larger than in comparable West European facilities.

One might hypothesize that lead times in the transfer of Western technology would be reduced as Soviet organisations learn more about inviting tenders and negotiating contracts in the West. But the period between initial inquiry and contract signing is not likely to be decreased much, because of the Soviet separation of foreign trade organisations and end-users, bureaucratic conservatism, and pursuit of favourable prices and credit terms to save convertible currency. In turn, the various problems responsible for the long interval between delivery of Western equipment and commissioning of plants incorporating it are persistent features of the Soviet economy also likely to continue.

2. Diffusion

Diffusion may be viewed as the final stage in the process of technology transfer, as it concerns the spread of new (in this case imported Western) technology from its initial application to other units in the economy. The speed and extent of diffusion show the economy's ability to replicate (and even to improve) the original technology and to use it for production in large volume at satisfactory cost and quality. Thus, diffusion is a key factor in technological progress – even more important than the earlier stages of R & D and innovation, according to some Western specialists on technological progress[56].

The rate of diffusion might be measured by the number or percentage of potential adopting units that have actually done so, but it is difficult to identify this population of potential adopting units from among all existing and new units in a branch or sub-branch of industry. Thus, more commonly the rate of diffusion is measured by the share of total production capacity, output, or perhaps employment attributable to the new product or process[57].

However, the published Soviet literature offers little information of this sort about the technology imported from the West in the 1970s. Western firms supplying (embodied and disembodied) technology to the USSR generally know even less about the diffusion phase than about the absorption phase of Soviet assimilation experience. Furthermore, because absorption and diffusion of a new Western product or process may easily take 5-10 years, evidence of successful diffusion of some Western technology imported in the mid-1970s would not be available until some time in the mid-1980s.

Nonetheless, because of the importance of diffusion for Soviet technological progress and thus economic growth, it is appropriate to examine some factors affecting the rate of diffusion of new technology in developed market economies that appear to be applicable also to a socialist planned economy like that of the USSR.

The Western technical literature identifies several categories of factors influencing the speed and extent of diffusion:

1. The rate of investment;
2. The nature of the innovation;
3. The characteristics of potential adopters; and
4. The amount and kind of information available about the innovation[58].

1. Most new technology is introduced and then spread through investment in plant and equipment, either through the construction of new facilities or through the renovation of existing facilities. Hence, a higher rate of new fixed investment provides more potential opportunities to diffuse new technology. But the extent to which new investment incorporates and diffuses new, rather than traditional, technology depends on various factors, discussed below. Also, if older facilities are scrapped, because of falling demand or the competition of other producers, the share of the new technology in total production capacity or output will be greater – and the rate of diffusion so measured will be higher.

2. In regard to the new technology itself, diffusion is likely to be greater *a)* the less complex or novel the innovation[59], *b)* the smaller the cost of acquisition and installation, *c)* the greater the effect in raising output and/or in reducing the use of key inputs (which in the particular instance may be labour, materials, or energy)[60], and *d)* the smaller the demands for related innovation by suppliers upstream and customers downstream.

3. In regard to potential adopters – in the Soviet case, central planning agencies and branch ministries, rather than enterprise managements – diffusion is likely to be greater when the technical capability to adapt foreign technology to domestic conditions is greater, and

when inputs of labour and materials required for the (adapted) foreign technology are more easily obtained.

4. The speed and extent of diffusion of a superior foreign technology will be greater the quicker, the more detailed, and the more widespread is the dissemination of information about its nature, about the requirements for and results of its adaptation and application in domestic conditions (i.e. experience in the absorption phase), and about the extent of its superiority over alternative domestic (and East European) technologies.

As noted above, information is lacking about the extent to which the USSR has attempted to diffuse Western technology imported since 1970, after absorbing it. At best, one may conjecture that diffusion would often be difficult – and thus sometimes not be tried, or if tried not succeed – for a number of reasons. These are suggested by the four sets of factors affecting the rate of diffusion just discussed, by the handicaps to technological progress analysed in Chapter 2, by the difficulties in copying foreign technology considered in Chapter 3, and by the problems in absorption examined in the preceding sub-section.

The growth rates of investment in fixed capital and of the increment to productive fixed assets began to decline in the USSR in the late 1960s. For example, the average annual rate of growth of new fixed investment is now 2-3 per cent, compared with 6.4 per cent in 1966-70. (See Tables 1 and 2 in Chapter 2.) Also, Soviet innovation tends to be confined to only a part of total investment in an industrial branch (like chemicals or iron and steel), with major shares of new investment still embodying traditional technologies. In contrast, in Western developed market economies new capacity is devoted entirely to new technologies. Finally, in the USSR diffusion is reduced by the Soviet practices of low depreciation rates and the tardy withdrawal of obsolete facilities from production.

On the one hand, the diffusion process in the USSR should be aided by the absence of the commercial secrecy of profit-seeking capitalist firms and by the ability of planning agencies to allocate funds and inputs to high-priority projects.

On the other hand, the centrally planned economy of the USSR lacks the domestic and foreign competitive pressures that stimulate the diffusion of new technology in market economies. Also in the USSR the diffusion process is hampered by the shortcomings in organisation, financing, pricing, performance indicators, and incentives that retard Soviet technological progress generally. Shortages of Soviet labour, materials, and equipment delay the construction of new facilities and curtail their output once they are completed. In many cases, diffusion would still require at least some Western machinery and equipment, entailing convertible currency expenditures. Finally, national security restrictions may hamper the diffusion of new technology from military to civilian uses.

The machinery, equipment, and production know-how in imported turnkey plants often constitute a highly integrated, essentially indivisible complex designed with particular processes and equipment for optimum performance in specific operations to make a specific product in a specific location. In these circumstances, it would not be appropriate to copy – and would frequently be difficult to adapt – a part of a turnkey plant for use in the different conditions of another facility. Replication of the entire turnkey plant in another location is even less likely.

Despite the assimilation problems analysed in this section, it is reasonable to expect that the USSR will – albeit slowly and imperfectly – absorb and then diffuse part, if not all, of the technology imported from the West. However, the impact on Soviet industry and the Soviet economy as a whole will be modest, although noticeable in some particular sub-branches, such as passenger cars and certain chemicals.

NOTES AND REFERENCES

1. In principle, one might try a) to revalue Soviet foreign trade figures into "domestic" rubles or b) to revalue Soviet output (or investment or capital stock) figures into world market prices. The results of the two approaches are likely to differ considerably. For example, using the former approach, one recent study estimated that in 1980 exports and imports represented, respectively, 6.9 and 20.1 per cent of Soviet net material product (roughly equivalent to GNP minus most services and depreciation). In contrast, an estimate by the second approach yielded ratios of 6 per cent for exports and 4 per cent for imports. The difference in the two sets of results is due chiefly to the relatively high Soviet domestic prices of manufactured goods compared to those of primary products. See Vladimir G. Treml and Barry L. Kostinsky, *The Domestic Value of Soviet Foreign Trade: Exports and Imports in the 1972 Input-Output Table* (Foreign Economic Report No. 20; US Department of Commerce, Bureau of the Census, 1982); "Dependence of the Soviet Economy on Foreign Trade", Wharton Econometric Forecasting Associates (WEFA), *Centrally Planned Economies Current Analysis,* 14th July 1982; and Vladimir G. Treml, "Measuring the Role of Foreign Trade in the Soviet Economy", WEFA, *Centrally Planned Economies Current Analysis,* 6th August 1982.
2. Hanson, *Trade,* p. 129.
3. *Ibid.,* pp. 144-146.
4. Cf. John P. Hardt, "The Role of Western Technology in Soviet Economic Plans", in *East-West Technological Co-operation,* pp. 318-319.
5. Hanson, *Trade,* pp. 221-222.
6. A recent study of six developed market economies concluded that imported technology did not displace domestic R & D, as complementarity was stronger than substitution in the relationship between them. See Tuvia Blumenfeld, "A Note on the Relationship between Domestic Research and Development and Imports of Technology", *Economic Development and Cultural Change,* Vol. 27, No. 2 (January 1979), pp. 303-306. However, this study concerns market economies, in which industrial R & D are commonly performed by large manufacturing enterprises and integrated with their production goals. The complementary relationship in these circumstances found by Blumenfeld may not apply to the same degree in the Soviet economy, where R & D and production are carried out by separate organisations (as explained in Chapter 2, Section A).
7. Stanislaw Gomulka and Alec Nove, "Contribution to Eastern Growth: An Econometric Evaluation", in *East-West Technology Transfer* (Paris: OECD, 1984), pp. 11-51. See also Hanson, Trade, pp. 146-155, and Brada, "Technology Transfer", pp. 260-263. The following summary draws upon all three, which may be consulted for a more detailed technical discussion of both the theoretical models and the econometric measurement problems.
8. Capital formation could also increase if a country's domestic capacity in the capital goods sector is fully utilised and if the country exports consumer goods and imports capital goods. The first but not the second condition may be applicable in the Soviet case.
9. Hanson, *Trade,* p. 129.
10. Padma Desai, "The Productivity of Foreign Resource Inflow into the Soviet Economy", *American Economic Review,* Vol. 69, No. 2 (May 1979), pp. 70-75, and "The Rate of Return on Foreign Capital Inflow to the Soviet Economy", in US Congress, Joint Economic Committee, *Soviet Economy in a Time of Change,* Vol. 2, pp. 396-413.
11. Stanislaw Gomulka and Alec Nove, *Econometric Evaluation of Technology Transfer,* working document, OECD, Paris, 1980, p. 13.
12. Donald W. Green and Herbert S. Levine, "Soviet Machinery Imports", *Survey,* Vol. 23, No. 2 (Spring 1977-1978), pp. 112-125, and Donald W. Green and Herbert S. Levine, "Macroeconomic

Evidence of the Value of Machinery Imports to the Soviet Union", in Thomas and Kruse-Vaucienne, *Soviet Science*, pp. 394-423, present revised estimates, different from those published earlier in Donald W. Green and Herbert S. Levine, "Implications of Technology Transfers for the USSR", in *East-West Technological Co-operation*, pp. 48-73.

13. For detailed discussion of these and other criticisms of the estimates of Green and Levine, see Martin L. Weitzman, "Technology Transfer to the USSR: An Econometric Analysis", *Journal of Comparative Economics*, Vol. 3, No. 2 (June 1979), pp. 145-166; Yasushi Toda, "Technology Transfer to the USSR: The Marginal Productivity Differential and the Elasticity of Intra-capital Substitution in Soviet Industry", *Journal of Comparative Economics*, Vol. 3, No. 2 (June 1979), pp. 181-194; Donald W. Green, "Technology Transfer to the USSR: A Reply", *Journal of Comparative Economics*, Vol. 3, No. 2 (June 1979), pp. 178-180; Gomulka and Nove, "Contribution", pp. 18-20; Hanson, *Trade*, pp. 146-153; and Josef C. Brada and Dennis L. Hoffman, "Technology Transfer to the USSR and the Shape of the Production Function", *De Economist* (Amsterdam), Vol. 130 (1982), No. 3, pp. 420-427.
14. Gomulka and Nove, *Econometric Evaluation*, pp. 22-25. Gomulka's model is explained in more detail in his "Growth and the Import of Technology: Poland, 1971-1980", *Cambridge Journal of Economics*, Vol. 2, No. 1 (March 1978), pp. 1-16.
15. Cf. Hanson, *Trade*, pp. 155 and 211.
16. For a survey of case studies covering various aspects of the transfer of Western technology to Eastern Europe as well as the USSR, see George D. Holliday, "Survey of Sectoral Case Studies", in *East-West Technology Transfer* (Paris: OECD, 1984), pp. 55-94.
17. Organisation for Economic Co-operation and Development (OECD), *East-West Trade in Chemicals* (Paris: OECD, 1980), p. 36.
18. Hanson, *Soviet Strategies*, p. 15.
19. US Central Intelligence Agency, *Soviet Chemical Equipment Purchases from the West: Impact on Production and Foreign Trade* (ER 78-10554; Washington, D.C., 1978), pp. 4-5 and 14-15.
20. *Ibid.*, pp. 5-6.
21. For lists of specific orders for Western chemical plant and equipment, see Hanson, *Soviet Strategies*, pp. 61-95, and US Central Intelligence Agency, *Soviet Chemical Equipment Purchases*, pp. 20-34.
22. US Central Intelligence Agency, *Soviet Chemical Equipment Purchases*, p. 8.
23. Vladimir Sushkov, "Foreign Trade for the Agro-Industrial Complex", *Foreign Trade* (Moscow), 1983, No. 1, p. 4.
24. *Ibid.*, pp. 9-11.
25. *Ibid.*, pp. 14-18.
26. For a more detailed explanation, see Hanson, *Trade*, pp. 161-185.
27. US Central Intelligence Agency, *USSR: Role of Foreign Technology in the Development of the Motor Vehicle Industry*, pp. 6-8.
28. *Ibid.*, p. 14.
29. Holliday, *Technology Transfer*, pp. 143-152.
30. Patrick Gutman, "East-West Industrial Co-operation in the Automobile Industry and the International Division of Labor", *The ACES Bulletin*, Vol. XXII, No. 3-4 (Fall-Winter 1980), p. 2. This article is a translation of his "Coopération industrielle Est-Ouest dans l'automobile et modalités d'insertion des pays de l'Est dans la division internationale du travail occidentale", *Revue d'Études Comparatives Est-Ouest*, Vol. XI, No. 2 (juin 1980), pp. 99-154, and No. 3 (septembre 1980), pp. 57-100.
31. *The Wall Street Journal*, 28th November 1983, p. 34.

32. US Central Intelligence Agency, *USSR: Role of Foreign Technology in the Development of the Motor Vehicle Industry,* p. 8.
33. Serge Schmemann, "Brezhnev's Souvenir: Vast, Limping Truck Factory on a Windswept Plain", *The New York Times,* 4th February 1983, p. 4.
34. Holliday, *Technology Transfer,* pp. 159-164.
35. *Ibid.,* pp. 147-150.
36. S. Bogatko and N. Morozov, "KamAZ: put' k sovershenstvu" [The Kama Automotive Factory: the road to improvement], *Pravda,* 9th July 1984, p. 3, and 10th July 1984, p. 2.
37. Philip Hanson and Malcolm Hill, "Soviet Assimilation of Western Technology: A Survey of UK Exporters' Experience", in US Congress, Joint Economic Committee, *Soviet Economy in a Time of Change,* Vol. 2, pp. 586-587.
38. Grant, "Soviet Machine Tools", pp. 573-576.
39. *Ibid.,* p. 578.
40. *Ibid.,* pp. 578-579. However, the USSR subsequently developed similar domestic grinding equipment. See Thane Gustafson, *Selling the Russians the Rope? Soviet Technology Policy and US Export Controls* (R-2649-ARPA; Santa Monica, Calif.: Rand Corporation, 1981), pp. 10-14.
41. Grant, "Soviet Machine Tools", pp. 578-579.
42. US Congress, Office of Technology Assessment, *Technology,* pp. 174-183.
43. The following discussion of drill pipe, drill bits, drilling rigs, submersible pumps, offshore equipment, oil refineries, field automation systems, and gas-lift technology is based on US Congress, Office of Technology Assessment, *Technology,* pp. 54-70.
44. This discussion of gas pipelines is based on Campbell, *Soviet Energy Technologies,* pp. 204-221, unless otherwise noted.
45. Gustafson, *The Soviet Gas Campaign,* pp. 89-91, and Ed. A. Hewett, "Near-Term Prospects for the Soviet Natural Gas Industry and the Implications for East-West Trade", in US Congress, Joint Economic Committee, *Soviet Economy in the 1980s,* Part 1, pp. 408-409.
46. However, the USSR has intensified efforts to increase domestic production of 16- and 25-megawatt turbines for compressors in response to US restrictions in 1981-1982 on the export of US oil and gas technology to the USSR. See Gustafson, *The Soviet Gas Campaign,* pp. 92-99, and Hewett, "Near-Term Prospects", pp. 406-407.
47. Hewett, "Near-Term Prospects", pp. 405 and 409. See Gustafson, *The Soviet Gas Campaign,* pp. 109-111, for a tabulation of Western sales of 56-inch gas pipe to the USSR in 1980-1982, compiled from Western press reports. See John B. Hannigan and Carl H. McMillan, *The Soviet-West European Energy Relationship: Implications of the Shift from Oil to Gas* (Research Report No. 20; Ottawa, Canada: Carleton University Institute of Soviet and East European Studies, 1983), pp. 76-79, for tabulations of Western contracts for pipe and equipment for the Urengoi pipeline network.
48. Hewett, "Near-Term Prospects", pp. 397-402, and Gustafson, *The Soviet Gas Campaign,* pp. 54-82.
49. Hannigan and McMillan, *The Soviet-West European Energy Relationship,* pp. 3-10.
50. Unless otherwise noted, this sub-section is based on Kathleen E. Braden, "The Role of Imported Technology in the Export Potential of Soviet Forest Products", in Robert G. Jensen, Theodore Shabad, and Arthur W. Wright (eds.), *Soviet Natural Resources in the World Economy* (Chicago: University of Chicago Press, 1983), pp. 442-463.
51. Malkevich, *East-West Economic Co-operation,* pp. 119-120.
52. For other recent examples, see "Siberian Pipeline Material Stranded in Leningrad Port", *The Japan Times,* 7th May 1983, p. 8; John F. Burns, "A Tough Time for Siberian Mining", *The New*

York Times, 22nd May 1983, Section 3, p. 9; and "Mnogolikaia bezotvetsvennost'" [Multifaceted irresponsibility], *Pravda,* 21st December 1983, p. 2.

53. See, for example, Smeliakov, "I spros", concerning diesel engines, and Iu. Arakelian, "Staritsia po litsenzii" [Aging under licence], *Pravda,* 10th March 1982, p. 3, regarding equipment to produce chloroprene rubber.

54. V.K. Fal'tsman, "Narodnokhoziaistvennyi zakaz na novuiu tekhniku" [National economic order for new equipment], *Ekonomika i organizatsiia promyshlennogo proizvodstva,* 1983, No. 7, p. 16.

55. The following composite account is based on the results of surveys of British firms in Hanson and Hill, "Soviet Assimilation", and Malcolm R. Hill, *East-West Trade, Industrial Co-operation, and Technology Transfer: The British Experience* (Aldershot, Eng.: Gower Publishing Co., 1983), pp. 49-74, and surveys of West German firms in Karl Ch. Röthlingshofer and Heinrich Vogel, *Soviet Absorption of Western Technology: Report on the Experience of West German Exporters* (Munich: IFO, 1979). For a more detailed review of these surveys, see Hanson, *Trade,* pp. 186-208.

56. Cf. G.F. Ray, "Introduction", in L. Nasbeth and G.F. Ray (eds.), *The Diffusion of New Industrial Processes* (Cambridge, Eng.: Cambridge University Press, 1974), pp.4-5.

57. Gerhard Rosegger, *The Economics of Production and Innovation: An Industrial Perspective* (Oxford: Pergamon Press, 1980), pp. 239-240.

58. *Ibid.,* pp. 247-255; L. Nasbeth, "Summary and Conclusions", in Nasbeth and Ray (eds.), *The Diffusion,* pp. 302-310; George F. Ray, "The Diffusion of Mature Technologies", *National Institute Economic Review,* No. 106 (December 1983), pp. 56-62; and George F. Ray, *The Diffusion of Mature Technologies* (Cambridge, Eng.: Cambridge University Press, 1984), pp. 76-91.

59. Anita Benvignati, "The Relationship between the Origin and Diffusion of Industrial Innovation", *Economica,* Vol. 49, No. 195 (August 1982), p. 315.

60. The management of a capitalist firm would assess these in terms of the effect on profitability and of the relationship between the rate of return and the cost of capital, while Soviet planning agencies would assess them by formulas and criteria for evaluating the "effectiveness" of new technology.

Chapter 5

THE IMPACT OF THE TRANSFER OF WESTERN TECHNOLOGY ON SOVIET FOREIGN TRADE

An appraisal of the impact of the transfer of Western technology on Soviet foreign trade requires an examination of both the commodity composition of Soviet foreign trade and its geographical distribution by major market areas. Section A of this chapter discusses the commodity composition and geographical distribution of Soviet foreign trade as a whole. Section B examines in detail the role of the transfer of Western technology in Soviet trade with Western developed market economies in the OECD. Sections C and D consider more briefly some implications of the transfer of Western technology to the USSR for its trade with Eastern Europe and with non-Communist less developed countries, respectively.

According to Soviet official statistics, machinery and equipment represent about one-third, and food products another one-fourth, of the USSR's total imports. Petroleum and petroleum products constitute more than one-third, and machinery and equipment about an eighth, of total Soviet exports. About two-fifths of the USSR's total foreign trade turnover (imports plus exports) is with the six East European centrally planned economies. Developed market economies account for about one-third, and less developed market economies about one-eighth, of total Soviet foreign trade turnover.

From developed market economies the USSR imports primarily machinery, equipment, metal products, and chemicals – chiefly for the Soviet engineering and metalworking, chemical, and energy industries – as well as cereals. Fuels constitute two-thirds – and petroleum and petroleum products alone more than half – of Soviet exports to developed market economies. In trade with these countries, some Soviet imports and exports are linked by compensation agreements – for instance, involving Soviet imports of pipe and compressors for natural gas pipelines and Soviet exports of natural gas. There are other examples of a close connection between Soviet imports of Western technology and subsequent Soviet exports to the West in the chemical and automotive industries. On the whole, however, there is little evidence of a strong competitive threat in Western, or third area, markets attributable to the transfer of Western technology to the USSR.

Mineral fuels constitute about half of total Soviet exports to Eastern Europe. Machinery and equipment account for about three-fifths of Soviet imports from Eastern Europe. There is no indication of large-scale transfer of Western technology to the USSR through Eastern Europe (rather than directly). But Western technology acquired by the USSR has contributed to Soviet exports to Eastern Europe of natural gas, motor vehicles, and other products.

In trade with less developed market economies, food, beverages, and tobacco dominate Soviet imports, while fuels, machinery and equipment, and arms are the main Soviet exports. Technology acquired from the West does not play any significant role in Soviet trade with less developed market economies.

A. AGGREGATE SOVIET FOREIGN TRADE

According to Soviet official foreign trade statistics, machinery and equipment constitute the most important single category of Soviet imports from the rest of the world. As Table 21 shows, this category has accounted for about one-third of total Soviet imports. Next in relative importance are food and non-food consumer goods. However, from 1970 to 1981 imports of food grew more rapidly (tenfold) than imports of machinery and equipment (fourfold). The percentage share of machinery rose to 34.4 in 1982 and 38.2 in 1983, while the percentage share of food declined to 23.7 in 1982 and 20.5 in 1983[1].

In turn, Table 22 presents data on the commodity composition of total Soviet exports. Most striking is the sharp rise in the percentage share of petroleum and petroleum products, from 11.5 in 1970 to 37.8 in 1981. Although the value of machinery and equipment exports grew threefold over the period, their percentage share in total exports dropped from 21.5 to 13.7. From 1970 to 1981 exports of chemicals increased by 554 per cent, but their share in total exports in 1981 was only 3.0 per cent. These trends continued through 1983, when the share of fuels in total exports reached 53.7 per cent, while that of machinery and equipment was 12.5 per cent[2].

Table 23 shows the distribution of Soviet foreign trade by major country groups that constitute four distinct market areas, for a combination of political and economic reasons. The group composed of the six centrally planned economies of Eastern Europe (Bulgaria, Czechoslovakia, the German Democratic Republic, Hungary, Poland, and Romania) is the Soviet Union's most important trading partner for both Soviet exports and Soviet imports. However, from 1970 to 1982 Eastern Europe's percentage share in total Soviet exports fell from 52.8 to 41.6, and the drop in its percentage share in total Soviet imports was from 56.5 to 43.1. In 1983 Eastern Europe accounted for 42.9 per cent of Soviet exports and 46.2 per cent of Soviet imports[3].

In contrast, the role of Western developed market economies in Soviet foreign trade rose markedly from 1970 to 1982. Their percentage share in total Soviet exports climbed from 19.2 to 30.1, and their percentage share in total Soviet imports grew from 24.0 to 33.7. In 1983, the developed market economies took 28.9 per cent of Soviet exports and supplied 31.4 per cent of Soviet imports[4].

The value of Soviet trade with non-Communist less developed countries increased sixfold from 1970 to 1982. But there was relatively little change in their percentage share in total Soviet exports (from 15.4 to 15.9) or in total Soviet imports (from 10.9 to 11.7). In 1983 these countries accounted for 15.5 per cent of Soviet exports and 12.0 per cent of Soviet imports[5].

However, Tables 21-23 are based on Soviet official foreign trade statistics in "foreign-trade rubles" that combine:

1. Soviet trade with developed and less developed non-Communist countries at world market prices (and chiefly in convertible currencies); and
2. Soviet trade with Communist countries at CMEA contract prices (and primarily through accounts in "transferable rubles" that are inconvertible and non-transferable).

The validity of this aggregation, and thus of calculations of shares of components in totals – though common in Soviet publications and also in many Western studies – may be questioned for several reasons.

First, CMEA contract prices differ from world market prices – and to different degrees for different products. CMEA contract prices are supposed to be based (with various

Table 21. **Soviet imports, by commodity group, 1979, 1976 and 1981** [a]

Commodity Group	1970		1976			1981		
	Value (Millions of US dollars)	% of total	Value (Millions of US dollars)	% of total	Index (1970=100)	Value (Millions of US dollars)	% of total	Index (1970=100)
Machinery and equipment	4 166	35.5	13 868	36.3	333	22 107	30.2	531
Petroleum and petroleum products	82	0.7	661	1.7	806	*b*	*b*	*b*
Coal and coke	124	1.1	480	1.3	387	*b*	*b*	*b*
Ores and concentrates	303	2.6	336	0.9	111	929	1.3	307
Ferrous metals and manufactures	593	5.1	3 029	7.9	511	4 574	6.3	771
Non-ferrous metals and manufactures	98	0.8	275	0.7	281	20	0.0 [c]	20
Chemicals	618	5.3	1 668	4.4	270	4 031	5.5	652
Rubber and rubber products	192	1.6	357	0.9	186	298	0.4	155
Wood and wood products	248	2.1	670	1.8	270	1 305	1.8	526
Textile raw materials and semi-manufactures	561	4.8	894	2.3	159	1 271	1.7	227
Food	1 590	13.6	7 736	20.2	487	18 371	25.1	1 155
Non-food consumer goods	2 199	18.8	4 838	12.7	220	9 161	12.5	417
Other	946	8.1	3 400	8.9	359	11 091	15.2	1 172
Total	11 720	100.0	38 212	100.0	326	73 158	100.0	624

a) Components may not add to totals because of rounding.
b) Not available.
c) Less than 0.1.
Source: Annex Table A-12.

Table 22. **Soviet exports, by commodity group, 1979, 1976 and 1981** [a]

Commodity Group	1970		1976			1981		
	Value (Millions of US dollars)	% of total	Value (Millions of US dollars)	% of total	Index (1970=100)	Value (Millions of US dollars)	% of total	Index (1970=100)
Machinery and equipment	2 753	21.5	7 219	19.4	262	10 862	13.7	395
Petroleum and petroleum products	1 469	11.5	10 210	27.4	695	30 029	37.8	2 044
Coal and coke	408	3.2	1 350	3.6	331	1 597	2.0	391
Ores and concentrates	403	3.2	839	2.3	208	1 016	1.3	252
Ferrous metals and manufactures	1 351	10.6	2 734	7.3	202	3 511	4.4	260
Non-ferrous metals and manufactures	627	4.9	1 138	3.1	181	b	b	b
Chemicals	364	2.8	939	2.5	258	2 380	3.0	654
Wood and wood products	831	6.5	1 995	5.4	240	2 632	3.3	317
Textile raw materials and semi-manufactures	437	3.4	1 067	2.9	244	1 545	1.9	354
Food	1 019	8.0	1 106	3.0	109	1 587	2.0	156
Non-food consumer goods	337	2.6	913	2.4	271	1 172	1.5	348
Others	2 788	21.9	7 759	20.8	278	23 046	29.0	827
Total	12 787	100.0	37 269	100.0	291	79 377	100.0	621

a) Components may not add to totals because of rounding.
b) Negligible.
Source: Annex Table A-12.

Table 23. **Value of Soviet exports and imports, by country group, 1970, 1976 and 1982[a]**
Millions of US dollars

Year		Total Trade	Communist Countries		Non-Communist countries	
			Eastern Europe	Other countries[b]	Developed countries	Less Developed countries
1970	Exports	12 787	6 752	1 607	2 453	1 975
	Imports	11 720	6 627	1 004	2 814	1 275
1976	Exports	37 269	17 432	4 458	10 269	5 109
	Imports	38 212	16 261	3 827	14 360	3 763
1982	Exports	87 170	36 288	10 821	26 224	13 837
	Imports	77 848	33 566	8 960	26 204	9 118

a) Components may not add to totals because of rounding.
b) Cuba, Mongolia, Yugoslavia, China, Korea and Communist Vietnam.
Source: Annex Table A-14.

adjustments) on a five-year moving average of previous world market prices. In a period of inflation on the world market, CMEA contract prices for many products would be below world market prices. As a result, the share of Soviet trade with Eastern Europe in total Soviet trade would be understated. Likewise, in statistics on the commodity composition of Soviet trade the percentage shares would be understated for goods traded relatively more heavily with Eastern Europe than with developed and less developed non-Communist countries.

But there is evidence that although some intra-CMEA trade, like that in oil, occurs at below-world-market prices, the reverse is true for some other goods, for instance some machinery whose inferior quality, compared to supposed Western counterparts, is not adequately recognised in markdowns from world market prices.

Also, conversions from foreign-trade rubles to US dollars are made at Soviet official exchange rates that overvalue the ruble.

Because relative price distortions in CMEA contract prices (compared to world market prices) vary considerably by commodity and by individual East European trading partner, there is no simple way to adjust Soviet official foreign trade statistics to a uniform valuation of flows in US dollars at a realistic exchange rate. Efforts to do so, in the absence of adequate data, involve many assumptions and have wide ranges of error. Hence, it is important to consider separately Soviet trade with different market areas involving different prices and different kinds of currency[6].

B. SOVIET TRADE WITH OECD COUNTRIES

Because published Soviet official statistics on the commodity composition of trade with different groups of countries are not sufficiently disaggregated, this discussion of Soviet trade with developed market economies is based on the much more detailed statistics available for OECD Member countries. These statistics refer to OECD countries' exports to the USSR, rather than Soviet imports from OECD countries, and to OECD countries' imports from the USSR, rather than Soviet exports to OECD countries. In principle in their mutual trade

OECD countries' exports should be equal to Soviet imports, and OECD countries' imports should be equal to Soviet exports. In practice two sets of "mirror" statistics will not be identical for various technical reasons – such as differences in the coverage of commodity categories, in the valuation of goods, and in the treatment of re-exports; and time in shipment that causes some goods dispatched by the exporter in one calendar period to be received by the importer in the next period[7]. However, these differences are of minor significance for this study.

In regard to Soviet trade with OECD countries, this section examines the commodity composition of Soviet imports and Soviet exports, and links between Soviet imports of Western technology and Soviet exports.

1. Imports

As Table 24 shows, manufactured goods – especially engineering products and iron and steel products – dominate Soviet imports from developed market economy countries. However, the share of food products, principally cereals, has risen sharply since 1970 as a result of heavy Soviet grain purchases in response to poor Soviet harvests. Thus, the percentage share of food in total Soviet imports from OECD countries reached 25.5 in 1982, compared with 3.8 in 1970.

Soviet imports from OECD countries particularly relevant to the transfer of Western technology were discussed in detail in Chapter 3, Section B, in the light of data in Tables 10 through 17 (based respectively on Annex Tables A-3 through A-10). Only the main findings of that analysis are summarised briefly here.

By *branch of origin,* machinery, equipment, and metal products; metallurgy; and chemicals dominate Soviet imports of *industrial goods* from OECD countries. Within the category of machinery, equipment, and metal products, the most important sub-branches are metal and woodworking machinery, ships and motor vehicles, construction and mining machinery, and scientific, measuring, and control equipment.

Table 24. **Commodity composition of OECD exports to the USSR, 1970, 1976, and 1982**[a]

Per Cent

Commodity Group	1970	1976	1982
Food	3.8	20.2	25.5
Cereals	3.3	15.7	17.2
Agricultural raw materials	6.2	3.0	2.7
Ores and other minerals	0.3	0.3	0.9
Fuels	0.2	0.3	1.3
Non-ferrous metals	1.0	0.5	0.8
Total primary products	16.5	24.4	31.3
Iron and steel	11.1	20.0	17.8
Chemicals	9.8	6.6	8.4
Engineering products	39.7	33.9	26.1
Road motor vehicles	1.1	2.9	2.7
Textiles and clothing	10.4	3.7	4.0
Other manufactures	10.8	7.7	8.8
Total manufactures	84.0	74.9	67.8
Residual	0.5	0.7	0.9
Total exports	100.0	100.0	100.0

a) Components may not add to totals because of rounding.
Source: OECD Data Bank

In regard to Soviet imports from OECD countries of *capital goods by type of product,* the largest and fastest-growing category (one-third of the total in 1982) is commodities for liquid fuel, gas, and water distribution – mainly tubes and pipes. Next in importance are ships, construction and mining machinery, and machine tools.

Analysis of the distribution of the total value of Soviet imports of *capital goods by end use* shows that in 1982 about one-half went to Soviet industry and about one-fourth to oil and gas distribution. Of the amount for industry in 1982, engineering and metalworking received about one-sixth, and mining and fuel extraction machinery another one-sixth, although these figures understate imports for the chemical industry, some of which are in the rather large "non-attributable" category.

Finally, in Soviet imports from OECD countries of *technology-based intermediate goods,* the most important categories are iron and steel, chemicals, and plastics.

In short, Soviet imports of manufactured goods from the OECD countries have been concentrated in machinery, equipment, metal products, and chemicals, and they have been destined primarily for the Soviet engineering and metalworking, chemical, and energy industries.

Additional information on the transfer of Western technology to the USSR is provided by the effort of Drabek and Slater to identify the technological sophistication of goods in Soviet trade with OECD countries. This approach involves grouping goods by their "R & D intensity" – the relative importance of R & D in their production – on the basis of classification schemes developed by the US Department of Commerce and the OECD. The results of any such exercise are, of course, very sensitive to both the classification scheme adopted and the data available to apply it, as Drabek and Slater acknowledge[8].

Tables 25 and 26 show, respectively in current dollars and in 1970 dollars, the distribution by broad demand category of OECD countries' exports to the USSR deemed to be "technology-based". Capital goods and intermediate goods each constitute about half of the total in current prices (Table 25). However, the share of capital goods is lower, and that of intermediate goods higher, at constant prices (Table 26), because the rate of inflation was much faster for capital goods than for intermediate goods. The shares of consumer goods and passenger cars are negligible in both tables.

In Table 27 these OECD countries' exports to the USSR of technology-based goods are classified by three categories of relative R & D intensity: "highly intensive", "moderately intensive", and "non-intensive". For the last category the term "non-intensive" is used in these statistics, but "low-intensity" would more accurately describe the commodities in the group. Exports of low R & D intensity (according to Western standards) represent more than four-fifths of total technology-based exports from the OECD countries to the USSR.

Further, as Table 28 shows, the share of technology-based products, including all three R & D intensity categories, in total exports of OECD countries to the USSR was only 57 per cent in 1982 (below a high of 73 per cent in 1970 and 1974). The rest of OECD countries' exports to the USSR, 43 per cent of the total in 1982, consisted of products judged not to be "technology-based". The combined share of goods of "high" and "moderate" R & D intensity in the OECD countries' total exports to the USSR (the first two rows in Table 28) ranged only from 7 to 14 per cent during 1970-1982[9].

2. Exports

For reasons explained earlier in the discussion of "mirror" statistics, Soviet exports to the OECD countries may be examined most conveniently through OECD statistics for imports of OECD countries from the USSR.

Table 25. **OECD exports of technology and technology-based products to the USSR, by broad demand category, 1970-1982**[a]

Demand category	1970	1971	1972	1973	1974	1975	1976	1977	1978	1979	1980	1981	1982
	Value (Millions of US dollars)												
Capital goods	1 132.3	927.3	1 227.4	1 892.6	2 520.3	4 934.7	5 672.3	6 016.1	6 538.9	6 559.0	5 723.4	5 623.9	7 275.6
Intermediate goods	800.7	975.7	1 187.2	1 561.3	3 276.9	4 231.7	3 938.6	4 091.8	4 685.2	6 071.1	7 614.1	6 706.2	6 571.7
Consumer goods	11.8	15.6	14.6	13.3	22.4	36.2	30.9	33.9	50.4	77.4	145.5	181.4	109.1
Passenger cars	1.8	1.5	2.8	3.6	4.6	4.5	2.6	5.0	4.5	5.1	5.0	4.8	6.1
Total	1 946.6	1 920.1	2 432.0	3 470.8	5 824.3	9 207.1	9 644.5	10 146.9	11 279.0	12 712.7	13 488.0	12 516.3	13 962.4
	Share (Per Cent)												
Capital goods	58.2	48.3	50.5	54.5	43.3	53.6	58.8	59.3	58.0	58.0	42.4	44.9	52.1
Intermediate goods	41.1	50.8	48.8	45.0	56.3	46.0	40.8	40.3	41.5	41.5	56.5	53.6	47.1
Consumer goods	0.6	0.8	0.6	0.4	0.4	0.4	0.3	0.3	0.4	0.4	1.1	1.4	0.8
Passenger cars	0.1	0.1	0.1	0.1	0.1	0.1	0.1	0.1	0.1	0.1	0.1	0.1	0.0
Total	100.0	100.0	100.0	100.0	100.0	100.0	100.0	100.0	100.0	100.0	100.0	100.0	100.0

a) Components may not add to totals because of rounding.
Source: East-West Technology Transfer Data Base, OECD, Paris.

Table 26. **Deflated value of OECD exports of tech[illegible]ogy and technology-based products to the USSR, by broad demand category, 1970-1981**[a]

Demand category	1970	1971	1972	1973	1974	1975	1976	1977	1978	1979	1980	1981
	Value (Millions of 1970 US dollars)											
Capital goods	1 132.3	835.2	972.8	1 201.7	1 367.3	2 302.9	2 686.5	2 489.9	2 552.4	2 162.3	1 792.5	2 141.9
Intermediate goods	800.7	1 006.0	1 155.1	1 225.7	1 727.5	1 938.5	2 241.6	2 000.4	2 580.0	2 485.4	2 993.4	3 098.5
Consumer goods	11.8	14.5	15.4	11.7	16.7	29.5	29.6	23.8	30.6	40.7	75.1	118.7
Passenger cars	1.8	1.3	2.2	2.2	2.6	2.2	1.2	2.1	1.6	1.6	1.4	1.6
Total	1 946.6	1 857.1	2 145.5	2 441.4	3 114.1	4 273.1	4 958.9	4 516.1	5 164.5	4 690.0	4 862.4	5 360.7
	Share (Per Cent)											
Capital goods	58.2	45.0	45.3	49.2	43.9	53.9	54.2	55.1	49.4	46.1	36.9	40.0
Intermediate goods	41.1	54.2	53.8	50.2	55.5	45.4	45.2	44.3	50.0	53.0	61.6	57.8
Consumer goods	0.6	0.7	0.7	0.5	0.5	0.7	0.6	0.5	0.6	0.9	1.5	2.2
Passenger cars	0.1	0.1	0.1	0.1	0.1	0.0	0.0	0.0	0.0	0.0	0.0	0.0
Total	100.0	100.0	100.0	100.0	100.0	100.0	100.0	100.0	100.0	100.0	100.0	100.0

a) Components may not add to totals because of rounding.
Source: East-West Technology Transfer Data Base, OECD, Paris.

Table 27. **Distribution of OECD exports of technology and technology-based products to the USSR, by R&D intensity category, 1970-1982[a]**

R&D Intensity Category	1970	1971	1972	1973	1974	1975	1976	1977	1978	1979	1980	1981	1982
	Value (Millions of U.S. dollars)												
Highly R&D intensive	38.0	34.5	42.2	51.5	71.3	110.3	106.0	122.7	160.5	193.9	198.1	184.8	194.4
Moderately R&D intensive	308.4	307.1	352.8	382.3	832.0	1 153.7	1 058.1	1 663.3	1 610.8	1 876.1	2 478.8	2 133.2	2 137.2
Non R&D intensive	1 599.8	1 460.4	1 855.8	2 767.7	4 604.9	7 158.3	7 853.1	7 849.7	8 970.5	10 006.2	10 268.3	9 666.1	10 870.9
Total	1 946.3	1 802.0	2 250.8	3 201.5	5 508.3	8 422.3	9 017.1	9 635.8	10 741.8	12 076.2	12 945.1	11 984.1	13 202.5
	Share (Per Cent)												
Highly R&D intensive	1.95	1.91	1.87	1.61	1.30	1.31	1.18	1.27	1.49	1.61	1.53	1.54	1.47
Moderately R&D intensive	15.85	17.04	15.67	11.94	15.11	13.70	11.73	17.26	15.00	15.54	19.15	17.80	16.19
Non R&D intensive	82.20	81.04	82.45	86.45	83.60	84.99	87.09	81.46	83.51	82.86	79.32	80.66	82.34
Total	100.00	100.00	100.00	100.00	100.00	100.00	100.00	100.00	100.00	100.00	100.00	100.00	100.00

a) Components may not add to totals because of rounding.
Source: East-West Technology Transfer Data Base, OECD, Paris.

Table 28. **Share of technology and technology-based products in OECD exports to the USSR, 1970.1982**

Per Cent

Category	1970	1971	1972	1973	1974	1975	1976	1977	1978	1979	1980	1981	1982
Technology and technology-based products													
Highly R&D intensive	1.44	1.30	1.08	0.89	0.94	0.88	0.76	0.89	1.03	1.01	0.92	0.83	0.84
Moderately R&D intensive	11.75	11.62	9.00	6.58	10.98	9.20	7.64	12.11	10.38	9.79	11.48	9.61	9.27
Non R&D intensive	59.73	55.26	47.34	47.64	60.76	57.05	56.68	57.15	57.83	52.20	47.57	43.56	47.15
Sub-total	72.92	65.18	57.42	55.11	72.68	67.13	65.08	70.15	69.24	63.00	59.97	54.00	57.26
Other products	27.08	31.82	42.58	44.89	27.32	32.87	34.92	29.85	30.76	37.00	40.03	46.00	42.74
Total exports	100.00	100.00	100.00	100.00	100.00	100.00	100.00	100.0	100.0	100.00	100.0	100.0	100.00

Source: East-West Technology Transfer Data Base, OECD, Paris.

Data on the commodity structure of these imports are presented in Table 29. Primary products have accounted for over four-fifths of the total value. Fuels alone constituted more than three-fourths of the total in 1982, compared with one-third in 1970. The chief reason was a sharp increase in the value of Soviet petroleum exports, mainly due to price increases rather than quantity growth. Although the volume of Soviet oil shipments to OECD countries is no longer increasing, natural gas deliveries have been rising and are likely to continue to do so.

Table 29. **Commodity composition of OECD imports from the USSR, 1970, 1976 and 1982**[a]

Per Cent

Commodity group	1970	1976	1982
Food	7.2	2.2	0.9
Agricultural raw materials	23.8	15.8	6.9
Ores and other minerals	8.7	4.3	1.6
Fuels	33.3	56.7	79.1
Coal and coke	7.0	4.8	1.2
Petroleum and petroleum products	25.7	48.1	62.0
Gas, natural and manufactured	0.6	3.8	15.6
Non-ferrous metals	11.4	4.1	2.4
Total primary products	84.4	83.1	91.0
Iron and steel	3.8	1.1	0.6
Chemicals	3.1	3.6	3.7
Engineering products	3.6	3.1	0.9
Road motor vehicles	0.3	1.3	0.9
Textiles and clothing	0.6	0.5	0.2
Other manufactures	3.3	6.6	2.1
Total manufactures	14.7	16.2	8.4
Residual	0.9	0.7	0.6
Total imports	100.0	100.0	100.0

a) Components may not add to totals because of rounding.
Source: OECD Data Bank.

An alternative classification of OECD countries' imports from the USSR by *sector of origin,* at current prices, appears in Table 30. It shows that the value of Soviet sales to OECD countries of products of industrial origin rose more than threefold from 1970 to 1976 and then doubled from 1976 to 1982. In contrast, the value of Soviet sales of products of agricultural origin in 1982 was slightly below the 1976 level. Similar data at constant 1970 prices in Table 31 (see also Annex Tables A-15 and A-16) reveal that price increases accounted for most of the growth since 1970 in the value of Soviet sales of both agricultural and industrial products to the OECD countries.

Tables 32 and 33, respectively, disaggregate by *branch of origin* the sales *of industrial goods* in Tables 30 and 31. (In turn, Annex Tables A-17 and A-18 disaggregate further the series in Tables 32 and 33.) The most striking development since 1970 is the growth in the value of Soviet sales of petroleum and gas (included in "products of mining") from $315 million in 1970 to $9 781 million in 1982, and the parallel growth of Soviet sales of gasoline, oils, and other petroleum products (included in "chemicals") from $312 million in 1970 to $8 242 million in 1982. But most of the value growth in these categories was due to price increases, rather than quantity increments. In real terms, from 1970 to 1981 the USSR

Table 30. **OECD imports from the USSR, by sector of origin, 1979, 1976 and 1982**[a]

Sector of Origin	1970		1976			1982		
	Value (Millions of US dollars)	% of total	Value (Millions of US dollars)	% of total	Index (1970 = 100)	Value (Millions of US dollars)	% of total	Index (1970 = 100)
Agricultural products								
Cereals	39.9	1.6	0.7	0.0	2	0.1	0.0	0
Fruits and vegetables	5.9	0.2	3.0	0.0	51	7.3	0.0	124
Inedible crude materials	50.4	2.0	86.4	0.8	171	90.3	0.3	179
Wood and lumber	278.0	10.9	597.1	5.5	215	564.7	2.2	203
Textiles fibres	57.4	2.2	441.1	4.1	768	454.7	1.8	792
Live animals	12.6	0.5	46.5	0.4	369	76.2	0.3	605
Tobacco (unmanufactured)	0.8	0.0	2.7	0.0	337	1.4	0.0	175
Other products	37.8	1.5	104.6	1.0	277	74.1	0.3	196
Total	482.9	18.9	1 282.1	11.9	266	1 268.8	5.1	263
Industrial products	2 071.0	81.1	9 483.4	88.2	458	23 791.1	94.9	1 149
Grand total	2 553.9	100.0	10 765.5	100.0	422	25 059.9	100.0	981

a) Components may not add to totals because of rounding.
Source: Annex Table A-15.

Table 31. **Deflated value of OECD imports from the USSR, by sector of origin, 1970, 1976 and 1981[a]**

Sector of Origin	1970		1976			1981		
	Value (Millions of 1970 US dollars)	% of total	Value (Millions of 1970 US dollars)	% of total	Index (1970 = 100)	Value (Millions of 1970 US dollars)	% of total	Index (1970 = 100)
Agricultural products								
Cereals	39.9	1.6	0.3	0.0	1	0.0	0.0	0
Fruits and vegetables	5.9	0.2	1.9	0.0	32	4.7	0.1	80
Inedible crude materials	50.4	2.0	36.5	0.8	72	55.1	1.3	109
Wood and lumber	278.0	10.9	280.2	6.0	101	202.5	4.7	73
Textile fibres	57.4	2.2	232.7	5.0	405	181.3	4.2	316
Live animals	12.6	0.5	26.0	0.6	206	22.0	0.5	175
Tobacco (manufactured)	0.8	0.0	1.2	0.0	150	1.7	0.0	213
Other products	37.8	1.5	70.4	1.5	186	41.9	1.0	111
Total	482.9	18.9	649.1	14.0	134	509.1	11.8	105
Industrial products	2 071.0	81.1	3 991.5	86.0	193	3 807.3	88.2	184
Grand total	2 553.9	100.0	4 640.6	100.0	182	4 316.5	100.0	169

a) Components may not add to totals because of rounding.
Source: Annex Table A-16.

Table 32. OECD industrial goods imports from the USSR, by branch of origin, 1970, 1976 and 1982[a]

Branch of Origin	1970		1976			1982		
	Value (Millions of US dollars)	% of total	Value (Millions of US dollars)	% of total	Index (1970 = 100)	Value (Millions of US dollars)	% of total	Index (1970 = 100)
Products of mining	587.1	28.3	3 377.0	35.6	575	10 083.4	42.4	1 717
Coke and manufactured natural gas	34.6	1.7	64.0	0.7	185	84.8	0.4	245
Energy	1.4	0.1	6.0	0.1	429	67.0	0.3	4 786
Food processing	124.1	6.0	187.2	2.0	151	169.1	0.7	136
Textiles	21.3	1.0	78.7	0.8	369	59.2	0.2	278
Clothing	0.7	0.0	1.2	0.0	171	1.8	0.0	257
Leather, shoes and furs	8.6	0.4	20.1	0.2	234	11.2	0.0	130
Paper, pulp and processed wood	241.7	11.7	592.5	6.2	245	611.2	2.6	253
Products of printing industry	0.6	0.0	2.6	0.0	433	5.3	0.0	883
Glass and china	3.1	0.1	9.6	0.1	310	6.8	0.0	219
Chemicals	385.7	18.6	2 884.9	30.4	748	9 153.4	3.8	2 373
Other non-metallic mineral products	4.1	0.2	4.7	0.0	115	3.7	0.0	90
Metallurgy	399.1	19.3	584.0	6.2	146	794.4	3.3	199
Machinery, equipment and metal products	90.4	4.4	474.5	5.0	525	421.1	1.8	466
Other industrial commodities	168.5	8.1	1 196.7	12.6	710	2 318.5	9.7	1 376
Total	2 071.0	100.0	9 483.4	100.0	458	23 791.1	100.0	1 149

a) The classification corresponds to the USSR branch of origin classification. Components may not add to totals because of rounding.
Source: Annex Table A-17.

Table 33. **Deflated value of OECD industrial goods imports from the USSR, by branch of origin, 1970, 1976 and 1981**[a]

Branch of Origin	1970		1976			1981		
	Value (Millions of 1970 US dollars)	% of total	Value (Millions of 1970 US dollars)	% of total	Index (1970 = 100)	Value (Millions of 1970 US dollars)	% of total	Index (1970 = 100)
Products of mining	587.1	28.3	859.0	21.5	146	735.3	19.3	125
Coke and manufactured natural gas	34.6	1.7	21.3	0.5	62	19.2	0.5	55
Energy	1.4	0.1	2.9	0.1	207	7.1	0.2	507
Food processing	124.1	6.0	116.6	2.9	94	82.6	2.2	67
Textiles	21.3	1.0	37.3	1.0	175	22.0	0.6	103
Clothing	0.7	0.0	0.6	0.0	86	0.8	0.0	114
Leather, shoes and furs	8.6	0.4	11.3	0.3	131	8.4	0.2	98
Paper, pulp and processed wood	241.7	11.7	280.4	7.0	116	217.9	5.7	90
Products of printing industry	0.6	0.0	1.2	0.0	200	1.9	0.0	317
Glass and china	3.1	0.1	13.4	0.3	432	4.5	0.1	145
Chemicals	385.7	18.6	712.2	17.8	185	827.2	21.7	214
Other non-metallic mineral products	4.1	0.2	6.5	0.2	159	1.4	0.0	34
Metallurgy	399.1	19.3	657.6	16.5	165	441.8	11.6	111
Machinery, equipment and metal products	90.4	4.4	243.5	6.1	269	200.8	5.3	222
Other industrial commodities	168.5	8.1	1 027.7	25.7	610	1 236.3	32.5	734
Total	2 071.0	100.0	3 991.5	100.0	193	3 807.3	100.0	184

a) The classification corresponds to the USSR branch of origin clasification. Components may not add to totals because of rounding.
Source: Annex Table A-18.

raised deliveries to the OECD countries of petroleum and gas by 100 per cent and of gasoline, oils, and other petroleum products by 45 per cent (see Annex Table A-18).

Table 34 provides a more detailed view of Soviet exports of *capital goods* to OECD countries. They amounted to only $174 million in 1982, down from $312 million in 1976, and a high of $374 million in 1979 (see Annex Table A-19).

OECD countries' imports from the USSR of *technology-based* intermediate goods, by type of product, are shown in Table 35. Organic and inorganic chemical elements have displaced non-ferrous metals as the most important category. Their respective percentage shares in the total value of sales were 42.9 and 31.2 in 1982, compared with 58.3 and 9.9 in 1970.

The technological level of Soviet exports to OECD countries has been analysed in the same way as the technological level of Soviet imports from OECD countries discussed in the preceding sub-section.

Tables 36 and 37 present data, at current prices and constant prices respectively, on OECD countries' imports of "technology-based" products from the USSR by broad demand category. The bulk of these are intermediate goods, representing a little more than three-fourths of the total. Passenger car sales grew from 1 to 9 per cent of the total in nominal terms and from 1 to 7 per cent in real terms.

These sales of "technology-based" goods are evaluated according to "R & D intensity" in Table 38. The share of goods judged "highly R & D intensive" was negligible. Those in the "non-intensive" (i.e. "low intensity") category accounted for 87.2 per cent of the total in 1970, but their share fell to 59.1 per cent by 1982, as the share of the "moderately intensive" category rose from 10.9 per cent to 38.8 per cent (below the peak of 41.3 per cent in 1979), due largely to the growth of Soviet exports of chemicals classified in this group.

In Table 39 the shares of Soviet sales in each "R & D intensity" category in total Soviet sales to OECD countries of "technology-based" and "non-technology-based" goods are given. These data indicate that the percentage share of "technology-based" exports from the USSR to OECD countries fell from 22.8 in 1970 to 9.32 in 1982.

A comparison of Table 28 (in the preceding sub-section) and Table 39 thus shows that in 1982 "technology-based" goods accounted for more than half of OECD countries' exports to the USSR but only one-tenth of Soviet exports to OECD countries. This finding is consistent with the Soviet technological lag discussed in Chapter 2 and the multifaceted transfer of Western technology to the USSR analysed in Chapter 3.

3. Relationship of Imports of Western Technology and Soviet Exports

With the available data it is not possible to obtain satisfactory measures on an aggregate (or sector or branch) basis of the impact of the transfer of Western technology on Soviet exports to developed market economies (or other groups of countries).

In order to establish relationships between an earlier transfer of Western technology and a subsequent increase in Soviet exports to the West one must know, or at least be able to make reliable assumptions about:

1. The speed and effectiveness of Soviet assimilation of Western technology;
2. The (assimilated) Western technology's contribution to output growth (stemming, in the case of embodied technology, from its share in the capital stock and its marginal productivity); and
3. The relationship between the growth of Soviet output and the growth of Soviet exports to the West.

Table 34. **OECD capital goods imports from the USSR, by type of product, 1970, 1976 and 1982**[a]

Type of product	1970		1976			1982		
	Value (Millions of US dollars)	% of total	Value (Millions of US dollars)	% of total	Index (1970 = 100)	Value (Millions of US dollars)	% of total	Index (1970 = 100)
Stationary power plants and water engineering	5.4	6.9	46.3	14.8	857	40.3	23.2	746
Electric power distribution	1.1	1.4	2.6	0.8	236	1.9	0.1	173
Liquid fuel, gas and water distribution	5.1	6.6	23.1	7.4	453	14.5	8.3	284
Transportation equipment	18.9	24.3	79.3	25.4	420	17.7	10.2	94
Agricultural equipment	3.4	4.4	27.9	8.9	821	13.5	7.8	397
Construction and mining machinery	4.3	5.5	12.5	4.0	291	10.1	5.8	235
Engineering, welding and metallurgical equipment (excluding furnaces)	18.6	23.9	52.4	16.8	282	31.5	18.1	169
Industrial and laboratory furnaces and gas generators	0.8	1.0	0.7	0.2	88	4.7	2.7	588
Machine and hand tools for working minerals, wood, plastics, etc.	0.2	0.3	0.6	0.2	300	0.4	0.2	200
Other electrical equipment	0.1	0.1	2.8	0.9	2 800	2.6	1.4	2 600
Electronics and telecommunications	2.9	3.7	9.4	3.0	324	12.7	7.3	438
Machinery for special industries	3.0	3.9	8.8	2.8	293	3.5	2.0	117
Mechanical handling equipment and storage tanks	4.5	5.8	7.5	2.4	167	5.6	3.2	124
Office machines	0.0[b]	0.0	0.9	0.3	[c]	0.3	0.2	[c]
Medical apparatus, instruments and furniture	0.3	0.4	0.9	0.3	300	1.5	0.9	500
Heating and cooling of buildings and vehicles	0.0	0.0	0.1	0.0	[c]	1.0	0.6	[c]
Measuring, controlling and scientific instruments	1.7	2.2	4.6	1.5	271	6.0	3.4	353
Finished structural parts and structures	3.3	4.2	11.0	3.5	333	0.3	0.2	9
Other capital equipment	4.2	5.4	21.0	6.2	500	5.5	3.2	131
Total	77.7	100.0	312.3	100.0	402	173.6	100.0	223

a) a) Components may not add to totals because of rounding.
b) Less than 0.1 million US dollars.
c) Not applicable, as an index number cannot be computed with a zero denominator.
Source: Annex Table A-19.

Table 35. **OECD imports of technology-based intermediate goods from the USSR, by type of product, 1970, 1976 and 1982**[a]

Type of Product	1970		1976			1982		
	Value (Millions of US dollars)	% of total	Value (Millions of US dollars)	% of total	Index (1970 = 100)	Value (Millions of US dollars)	% of total	Index (1970 = 100)
Parts and accessories of machinery and equipment	10.0	2.1	20.9	2.1	209	22.4	1.2	224
Parts and accessories of transport equipment	6.3	1.3	25.7	2.6	408	36.2	1.9	575
Paper manufactures	7.5	1.6	34.3	3.5	457	33.5	1.7	447
Textile manufactures	1.3	0.3	2.7	0.3	208	4.5	0.2	346
Synthetic rubber	0.7	0.1	8.6	0.9	1 229	27.3	1.4	3 900
Synthetic fibres	0.4	0.1	2.1	0.2	525	2.7	0.1	675
Special manufactures of leather, rubber, wood, glass and minerals	0.1	0.0	0.8	0.1	800	3.4	0.2	3 400
Miscellaneous mineral manufactures	0.3	0.1	0.2	0.0	67	0.1	0.0	33
Iron and steel	95.2	19.9	91.0	9.2	96	147.7	7.7	155
Non-ferrous metals	278.8	58.3	420.9	42.6	151	600.7	31.2	215
Organic and inorganic chemical elements	47.2	9.9	289.8	29.3	614	825.6	42.9	1 749
Final chemical manufactures	0.4	0.1	2.2	0.2	550	1.3	0.0	325
Chemical and plastic materials	28.8	6.0	86.8	8.8	301	213.3	11.1	741
Other intermediate goods	0.8	0.2	1.7	0.2	213	7.6	0.4	950
Total	477.9	100.0	987.8	100.0	207	1 926.0	100.0	403

a) Components may not add to totals because of rounding.
Source: Annex Table A-20.

Table 36. **OECD imports of technology and technology-based products from the USSR, by broad demand category, 1970-1982**[a]

Demand category	1970	1971	1972	1973	1974	1975	1976	1977	1978	1979	1980	1981	1982
	Value (Millions of US dollars)												
Capital goods	77.8	120.0	155.2	220.8	171.7	253.1	312.6	193.8	212.9	374.3	352.1	274.7	173.8
Intermediate goods	472.8	432.6	576.9	996.6	1 328.1	897.9	985.0	1 349.4	1 620.9	2 459.2	2 572.1	1 968.2	1 920.9
Consumer goods	7.8	11.2	15.9	20.6	24.5	30.8	25.8	27.7	30.1	35.1	37.9	20.6	12.6
Passenger cars	5.4	8.5	21.6	46.9	48.3	100.4	127.5	142.3	188.5	283.0	219.0	183.1	207.7
Total	563.8	572.2	769.7	1 284.9	1 572.7	1 282.2	1 451.0	1 713.2	2 052.3	3 151.5	3 181.1	2 446.6	2 315.2
	Share (Per Cent)												
Capital goods	13.8	21.0	20.2	17.2	10.9	19.7	21.5	11.3	10.4	11.9	11.1	11.2	7.5
Intermediate goods	83.8	75.6	75.0	77.6	84.4	70.0	67.7	78.8	79.0	78.0	80.9	80.4	83.0
Consumer goods	1.4	2.0	2.1	1.6	1.6	2.4	1.8	1.6	1.5	1.1	1.1	0.8	0.5
Passenger cars	1.0	1.5	2.8	3.7	3.1	7.8	8.8	8.3	9.2	9.0	6.9	7.5	9.0
Total	100.0	100.0	100.0	100.0	100.0	100.0	100.0	100.0	100.0	100.0	100.0	100.0	100.0

a) Components may not add to totals because of rounding.
Source: East-West Technology Transfer Data Base, OECD, Paris.

Table 37. **Deflated value of OECD imports of technology and technology-based products from the USSR, by broad demand category, 1970-1981**[a]

Demand category	1970	1971	1972	1973	1974	1975	1976	1977	1978	1979	1980	1981
	Value (Millions of US dollars)											
Capital goods	77.8	110.0	131.7	152.8	106.0	131.7	161.4	93.6	87.7	156.3	124.2	102.3
Intermediate goods	472.8	562.5	918.9	1 231.8	1 139.4	845.3	882.5	1 021.7	1 070.7	1 131.2	1 016.5	862.1
Consumer goods	7.8	10.4	13.7	14.1	16.2	18.1	14.7	14.5	13.6	14.4	15.0	10.8
Passenger cars	5.4	7.7	17.6	32.7	30.0	52.6	64.2	66.7	75.6	100.3	73.8	73.4
Total	563.8	690.6	1 081.9	1 431.4	1 291.5	1 047.7	1 122.8	1 196.4	1 247.6	1 402.2	1 229.5	1 048.5
	Share (Per Cent)											
Capital goods	13.8	15.9	12.2	10.6	8.2	12.6	14.4	7.8	7.0	11.1	10.1	9.8
Intermediate goods	83.8	81.5	84.9	86.1	88.2	80.7	79.6	85.4	85.8	80.7	82.7	82.2
Consumer goods	1.4	1.5	1.3	1.0	1.3	1.7	1.3	1.2	1.1	1.0	1.2	1.0
Passenger cars	1.0	1.1	1.6	2.3	2.3	5.0	5.7	5.6	6.1	7.2	6.0	7.0
Total	100.0	100.0	100.0	100.0	100.0	100.0	100.0	100.0	100.0	100.0	100.0	100.0

a) Components may not add to totals because of rounding.
Source: East-West Technology Transfer Data Base, OECD, Paris.

Table 38. **Distribution of OECD imports of technology and technology-based products from the USSR, by R & D intensity category, 1970-1982[a]**

R&D Intensity Category	1970	1971	1972	1973	1974	1975	1976	1977	1978	1979	1980	1981	1982
	Value (Millions of US dollars)												
Highly R & D intensive	10.9	11.4	28.4	29.2	38.9	36.3	52.8	50.9	82.8	69.8	146.2	65.8	47.7
Moderately R & D intensive	63.6	75.6	96.3	134.9	232.2	271.7	359.7	658.2	847.3	1 321.1	1 230.9	1 006.0	906.2
Non R & D intensive	507.7	503.5	655.2	1 142.7	1 346.6	1 018.8	1 064.1	1 025.0	1 150.9	1 807.3	1 844.5	1 433.6	1 380.2
Total	582.2	590.5	779.8	1 306.8	1 617.7	1 326.8	1 476.7	1 734.1	2 081.0	3 198.2	3 221.6	2 505.4	2 334.1
	Share (Per Cent)												
Highly R & D intensive	1.87	1.94	3.64	2.23	2.41	2.74	3.58	2.94	3.98	2.18	4.54	2.63	2.04
Moderately R & D intensive	10.93	12.80	12.34	10.32	14.35	20.48	24.36	37.96	40.72	41.31	38.21	40.15	38.82
Non R & D intensive	87.20	85.26	84.02	87.45	83.24	76.79	72.06	59.11	55.30	56.51	57.26	57.22	59.13
Total	100.0	100.0	100.0	100.0	100.0	100.0	100.0	100.0	100.0	100.0	100.0	100.0	100.0

a) Components may not add to totals because of rounding.
Source: East-West Technology Transfer Data Base, OECD, Paris.

Table 39. **Share of technology and technology-based products in OECD imports from the USSR, 1970-1982**

Per Cent

Category	1970	1971	1972	1973	1974	1975	1976	1977	1978	1979	1980	1981	1982
Technology and technology-based products													
Highly R & D intensive	0.43	0.40	0.90	0.60	0.49	0.43	0.49	0.42	0.63	0.38	0.61	0.28	0.19
Moderately R & D intensive	2.49	2.66	3.05	2.76	2.93	3.20	3.34	5.43	6.47	7.12	5.17	4.23	3.62
Non R & D intensive	19.88	17.71	20.73	23.38	17.00	11.99	9.88	8.45	8.79	9.73	7.75	6.03	5.51
Sub-total	22.80	20.77	24.68	26.74	20.42	15.62	14.37	14.3	15.89	17.23	13.53	10.54	9.32
Other products	77.20	79.23	75.32	73.26	79.58	84.38	85.63	85.7	84.11	82.77	86.47	89.46	90.68
Total imports	100.00	100.00	100.00	100.00	100.00	100.00	100.00	100.00	100.00	100.00	100.00	100.00	100.00

Source: East-West Technology Transfer Data Base, OECD, Paris.

However, it is difficult to establish the contribution of Western technology to the growth of Soviet output, for reasons explained in Chapter 4.

It is also difficult to show the relationship between increases in Soviet production and increases in Soviet exports to the West. First, for some commodities all or most of the additional output due to imports of Western technology may be intended for Soviet domestic use, rather than for export. Second, for some commodities the USSR can offer to increase exports to the West by reducing supplies for domestic use or exports to other areas, for instance Eastern Europe. Third, sales to the West depend not only on Soviet offers but also on Western demand, which is influenced by both economic conditions and government policies in Western countries.

Hence, efforts to correlate statistically Soviet purchases of Western technology and the growth of Soviet exports to the West are hampered by several serious problems. One is the high "sectoral" level of aggregation at which imports of Western technology and Soviet exports are compared. Other problems concern the rough, often questionable, assumptions about the contribution of Western technology to output growth, time lags between imports of Western technology and related Soviet exports, and the importance of supply versus demand factors in determining Soviet exports.

One such approach at measuring a relationship between Soviet purchases of Western technology and Soviet exports to the West is the calculation of Spearman rank correlation coefficients. For the period 1970-1977, Drabek ranked 16 industrial end-use branches by their shares in capital goods imports from OECD countries and by their average rates of growth of exports to OECD countries. The Spearman coefficients would be +1 for perfect similarity, -1 for perfect dissimilarity, and zero for no relationship. Drabek's result for the USSR was 0.3364. In analogous but more complex calculations, Hanson obtained a Spearman rank correlation coefficient of 0.405 for the USSR. In view of the conceptual and statistical limitations explained above, these results must be interpreted with caution. Moreover, they do not purport to measure directly the contribution of Western technology to Soviet export performance. Instead, they aim only to show to what degree the same "branches" (in the classification scheme adopted) led in both imports of Western technology and growth of exports to OECD countries[10]. A relatively low coefficient of 0.3-0.4 could still be compatible with – from the Soviet standpoint – satisfactory effects of imports of Western technology on the growth of Soviet production and Soviet exports to the West (and/or other areas).

One can identify some product groups for which there are clear strong links between imports of Western technology and subsequent Soviet exports to the West. The most obvious such link occurs through compensation agreements. As explained in Chapter 2, Section C, this form of countertrade involves Soviet imports of machinery, equipment, and other forms of technology from the West largely on credit. Repayment is subsequently made (starting several years later) through the sale of "resultant" goods produced through the use of the Western technology (and other inputs). The USSR makes relatively little use of the counterpurchase form of countertrade involving deliveries of unrelated goods.

The statistical data available on Soviet compensation exports to the West refer to scheduled deliveries as estimated from press reports and, sometimes, additional information supplied by Western firms. However, estimates of the value of scheduled or actual compensation deliveries may be inaccurate for several reasons.

First, they may understate scheduled deliveries because relevant information is not available from the Soviet side or from Western firms, which are more willing to reveal what they sell than what they buy[11].

Second, these estimates may overstate deliveries because they assume that the Soviet facilities start production on schedule, that planned output levels are achieved, and that planned sales can be accomplished at the intended prices. Actual compensation (or other countertrade) imports by Western countries cannot be calculated from Western foreign trade statistics, because customs declarations do not indicate which goods move under countertrade. Finally, the USSR expects to continue exporting goods to Western markets initially entered under compensation agreements after those agreements expire[12].

The link between the transfer of Western technology to the USSR and subsequent Soviet exports to the West is clear in the case of natural gas. Almost all of Soviet exports of natural gas to Western Europe – amounting to over $3 billion in 1982 – are under compensation agreements through which the USSR earlier obtained from developed market economies large-diameter pipe and compressor equipment for pipelines from Soviet gasfields to Eastern and Western Europe[13].

According to recent estimates, the USSR borrowed $15-20 billion in the West to finance imports of pipe and equipment from OECD countries for the gas pipeline network from the Urengoi fields in Northern Siberia. As explained above in Chapter 4, Section B, Subsection 4, this network has five domestic lines and one, Urengoi No. 6, to Western Europe. The USSR is scheduled to repay the Western credits during 1985-1994. The Urengoi No. 6 line went into operation in 1984 and is supposed to reach full capacity, planned at 40 billion cubic meters of gas, in 1986 or 1987. However, Western Europe's probable demand for Soviet gas at that time is lower, perhaps 30 billion cubic meters. At likely prices in the late 1980s, that quantity may earn the USSR about $6 billion per year[14]. However, this is only an approximate estimate of Soviet export revenue from Urengoi No. 6, because of the considerable uncertainties about both prices and quantities of future Soviet gas exports to Western Europe[15].

As explained in Chapter 4, Section B, the contribution of Western technology to Soviet exports of natural gas is primarily in transportation rather than production. In contrast, in the case of chemicals, Western technology has helped expand Soviet output. The Soviet chemical industry has made large purchases of Western machinery and equipment, often for turnkey projects and frequently complemented by industrial co-operation agreements. The bulk of the Western technology was for the production of fertilizers and artificial fibres.

By the late 1970s the growth of Soviet chemical production was reflected in rising exports to the OECD countries, particularly of ammonia, but also of other products like carbamide, polyethylene, and acrylonitrile[16].

Further increases in Soviet chemical exports to the West in the 1980s are scheduled under compensation and barter agreements[17]. By the mid-1980s, the annual value of Soviet chemical exports under such agreements is scheduled to reach close to $1 billion. The estimated approximate percentage shares of different types of chemicals in this total are as follows: ammonia, 30; other inorganic chemicals (like potash and sulphur), 12; organic chemicals (such as benzene, methanol, caprolactam, etc.), 30; urea, 19; and plastics, 9[18].

However, the USSR may not succeed in earning scheduled export revenues from sales of chemicals to the West. Quantities may be less than anticipated because of lower operating rates of Soviet plants and weak demand in Western markets. The latter factor may also lead to lower prices than originally expected[19]. The USSR has been willing to offer lower-than-prevailing prices in Western markets for ammonia, methanol, xylenes, and other products[20]. To charges of "dumping"[21], the USSR replies that, compared with Western producers, its costs are lower because it has large-scale plants with the latest technology located close to rich raw material sources.[22].

Passenger cars are another instance of a close link between Soviet import of Western technology and Soviet exports to the West, although not under compensation agreements as in the cases of natural gas and chemicals. As explained in Chapter 4, Section B, Western technology played an important role in the development of Soviet production of both passenger cars and trucks. The USSR exports under the name "Lada" a version of the passenger car modelled on the FIAT-124 that is sold in the USSR as the "Zhiguli". It also exports the "Niva", a four-wheel-drive vehicle of the jeep or land-rover type that is of Soviet conception but based on Western (FIAT) technology. Of total Soviet car exports of 252 000 units in 1982, 118 700 (chiefly Ladas) were sent to the West[23]. Both the Lada and the Niva are sold in Western markets at about 75 per cent of the price of similar West European and Japanese vehicles. This differential reflects in part the Lada's now somewhat outdated design and in part aggressive Soviet pricing. The Soviet share of the Western car market is insignificant and is likely to remain so in the face of competition by Western producers[24].

The Soviet Union has also obtained Western technology for machine tools, especially numerically-controlled tools (see Chapter 4, Section B), and as a result it now exports numerically-controlled and other machine tools incorporating Western technology[25]. Because these Soviet tools are technologically less advanced than current Western models, Soviet sales to the West are limited[26]. In 1982 machine tools accounted for only $29 million of the $174 million of capital goods that the USSR exported to the West (see Annex Table A-19).

On the whole, there is little evidence of a strong competitive threat in Western markets by Soviet exports attributable to the transfer of Western technology to the USSR[27]. However, in this respect two categories of products can be distinguished.

As illustrated in the case of chemicals, such a threat is more likely insofar as one or more of the following conditions apply:

1. The Western firms selling technology to the USSR are different from the Western firms dealing in the products the USSR seeks to export (e.g. chemical plant contractors versus chemical firms).
2. The Soviet export product is a homogeneous commodity, like basic or intermediate chemicals, for which customer choice among suppliers is more sensitive to price.
3. Western market demand for the product is weak relative to supply, because of overcapacity in the industry or the recession phase of the business cycle.
4. Western producers are unsuccessful in obtaining "antidumping" or "market disruption" relief from their respective governments.

In contrast, competition from Soviet exports will be less serious under the following conditions:

1. Western suppliers of technology to the USSR also make the final products, and they include in turnkey plant and industrial co-operation agreements with the USSR market-allocation clauses forbidding Soviet exports to Western markets (and perhaps other areas). This is common in the case of machine tools, for example[28].
2. Technological progress – especially product innovation – is rapid in the West. In accordance with the "product-cycle theory", in these circumstances Western firms sell aging (sometimes 6-10 years old) technologies to potential rivals, in the expectation that Western firms will preserve their technological lead and their competitive advantage in their home and also third markets. This condition applies to Western manufacturers of machine tools, computers, and passenger cars, for instance.

3. Where product differentiation is strong and (actual or perceived) quality differences among rival products are often more influential than price differences in buyers' decisions – for example, for manufactured versus primary products – the USSR is less able to increase exports by charging less than Western firms.
4. As in the case of natural gas, Soviet deliveries are attractive to Western countries when domestic supplies are insufficient and the reliability of alternative foreign suppliers is questioned.

C. SOVIET TRADE WITH EASTERN EUROPE

The six East European centrally planned economies account for more than 40 per cent of total Soviet imports and of total Soviet exports, according to Soviet official foreign trade statistics. (On distortions in these statistics, see Section A above.)

As Table 40 shows, in 1982 machinery and equipment accounted for three-fifths of total Soviet imports from Eastern Europe, and other manufactured goods an additional one-fifth. In contrast, as Table 41 indicates, fuels constituted half of Soviet exports to Eastern Europe. Crude materials (excluding fuels) represented a sixth of the total, as did machinery and equipment.

Table 40. **Commodity composition of Soviet imports from Eastern Europe, 1970 and 1976-1982** [a]

Per Cent

Commodity Group	1970	1976	1977	1978	1979	1980	1981	1982
Food, beverages and tobacco (SITC 0 and 1)	10.7	11.0	11.3	9.1	9.2	9.9	10.8	11.4
Crude materials, excluding fuels (SITC 2 and 4)	2.6	1.9	1.7	1.4	1.5	1.7	1.7	1.7
Mineral fuels and related materials (SITC 3)	2.6	3.6	2.8	2.6	2.2	1.7	1.1	0.2
Sub-total, Primary products	15.9	16.5	15.8	13.1	12.9	13.3	13.6	13.3
Chemicals (SITC 5)	6.9	6.8	6.6	6.3	5.8	6.3	6.8	7.4
Machinery and transport equipment (SITC 7)	47.1	50.9	52.9	57.7	58.5	57.7	58.3	60.3
Other manufactured goods (SITC 6 and 8)	29.2	24.3	23.2	18.0	22.1	22.4	20.9	18.5
Sub-total, manufactures	83.2	82.0	82.7	82.0	86.4	86.4	86.0	86.2
Commodities and transactions not classified according to kind (SITC 9)	0.9	1.5	1.5	4.9	0.7	0.4	0.3	0.5
Total	100.0	100.0	100.0	100.0	100.0	100.0	100.0	100.0

a) Components may not add to totals because of rounding.
Source: Computed from United Nations trade statistics in millions of US dollars in the following sources: 1970, 1976-77: United Nations, *Yearbook of International Trade Statistics, 1981* (New York, 1982), Vol. 1, Special Table C. 1978-1979: United Nations, *Monthly Bulletin of Statistics*, Vol. 37, No. 5 (May 1983), Special Table E. 1980-1982: United Nations, *Monthly Bulletin of Statistics*, Vol. 38, No. 5 (May 1984), Special Table D.

In its trade with Eastern Europe, the Soviet Union is a heavy net importer of machinery and equipment, the main channel of embodied technology transfer[29]. For example, in Soviet-East European trade in machinery and equipment during 1981-85, Soviet imports were planned at 60 billion foreign-trade rubles and Soviet exports at 35 billion foreign-trade rubles[30]. In this trade in 1983, Soviet imports were about 15 billion foreign-trade rubles and Soviet exports about 6 billion foreign-trade rubles[31].

Table 41. **Commodity composition of Soviet exports to Eastern Europe, 1970 and 1976-1982** [a]

Per Cent

Commodity Group	1970	1976	1977	1978	1979	1980	1981	1982
Food, beverages and tobacco (SITC 0 and 1)	6.2	0.8	2.5	1.2	2.3	1.4	1.6	0.6
Crude materials, excluding fuels (SITC 2 and 4)	17.3	12.5	14.1	11.7	13.0	13.1	17.6	15.1
Mineral fuels and related materials (SITC 3)	15.0	28.2	31.6	33.5	37.6	41.0	45.7	50.2
Sub-total, Primary products	38.5	41.5	48.2	46.4	52.9	55.5	64.9	65.9
Chemicals (SITC 5)	2.3	1.8	1.8	1.9	2.1	2.3	2.9	2.7
Machinery and transport equipment (SITC 7)	21.1	24.4	24.5	24.3	22.5	21.2	16.9	15.1
Other manufactured goods (SITC 6 and 8)	23.9	20.0	18.2	17.5	15.5	14.8	12.8	15.0
Sub-total, Manufactures	47.3	46.2	44.5	43.7	40.1	38.3	32.6	32.8
Commodities and transactions not classified according to kind (SITC 9)	14.1	12.3	7.3	9.8	7.0	6.3	2.5	1.2
Total	100.0	100.0	100.0	100.0	100.0	100.0	100.0	100.0

a) Components may not add to totals because of rounding.
Source: Computed from United Nations trade statistics in millions of US dollars in the following sources: 1970, 1976-77: United Nations, *Yearbook of International Trade Statistics, 1981* (New York, 1982), Vol. 1, Special Table C. 1978-1979: United Nations, *Monthly Bulletin of Statistics*, Vol. 37, No. 5 (May 1983), Special Table E. 1980-1982: United Nations, *Monthly Bulletin of Statistics*, Vol. 38, No. 5 (May 1984), Special Table D.

Trends in Soviet imports of capital goods from Eastern Europe by end-use branch are shown in detail in Table 42. These imports were distributed widely across the Soviet economy. The branches with the largest shares in 1980, agriculture and electricity, received only 9.0 and 8.4 per cent of the total, respectively. Over the period from 1970 to 1980, total Soviet imports of capital goods from Eastern Europe rose 352 per cent, without any significant changes in the shares of individual end-use branches.

Imports from Eastern Europe account for over 60 per cent of total Soviet imports of machinery and equipment, according to Soviet official foreign trade statistics. For some categories of machinery and equipment, the East European percentage share in total Soviet imports is even higher – for instance, railway rolling stock, almost 100; materials-handling equipment, about 88; energy-generating and electrical equipment, almost 80; and equipment for the food industry, almost 70[32].

To some extent, Soviet imports of machinery and equipment from Eastern Europe may be explained by the technological superiority of East European products compared to Soviet products. But another important factor is CMEA specialisation and co-operation agreements, among producers at a similar technological level, intended to achieve economies of scale. However, on the whole the technological level of Soviet machinery and equipment imports from Eastern Europe is below that of machinery and equipment imported by the Soviet Union from the West[33].

There is little evidence of large-scale technology transfer from the West to the USSR through Eastern Europe, rather than directly. But Eastern Europe does play such an intermediary role in some cases, for example in the chemical industry, shipbuilding[34], and the production of some types of computer equipment[35].

Data on Soviet exports of capital goods to Eastern Europe by end-use branch are provided in Table 43. Their total value (at CMEA contract prices) rose 431 per cent from 1970 to 1980, but the increase was faster for the industrial sector (787 per cent), especially electricity, metallurgy, and building materials. In 1980, the industrial sector accounted for 36.1 per cent of the total, followed by the transport and agricultural sectors, with 33.3 and 14.7 per cent, respectively.

Table 42. **Soviet imports of capital goods from Eastern Europe, by end-use branch, 1970, 1976 and 1980** [a]

End-use Branch	1970		1976			1980		
	Value (Millions of US dollars)	% of total	Value (Millions of US dollars)	% of total	Index (1970 = 100)	Value (Millions of US dollars)	% of total	Index (1970 = 100)
Electricity	178.09	6.5	503.64	7.7	283	1 045.20	8.4	587
Fuel and mining	48.24	1.8	101.95	1.6	211	451.32	3.6	936
Metallurgy	116.03	4.2	93.75	1.4	81	335.37	2.7	289
Engineering and metalworking	156.20	5.7	383.52	5.9	246	729.84	5.9	467
Chemicals	140.94	5.1	415.62	6.3	295	670.12	5.4	475
Building materials	23.32	0.8	82.98	1.3	356	123.23	1.0	528
Wood and paper	20.54	0.7	28.77	0.4	140	73.00	0.6	355
Textile	0.00[b]	0.0	0.00	0.0	100	0.00	0.0	100
Other light industry	105.57	3.8	302.85	4.6	287	659.85	5.3	625
Food and food processing	115.64	4.2	177.96	2.7	154	443.70	3.6	384
Total industry	904.57	32.9	2 091.04	31.9	231	4 531.64	36.5	501
Construction	49.76	1.8	39.55	0.6	79	169.63	1.4	341
Agriculture, including tractors	165.87	6.0	592.55	9.0	357	1 118.55	9.0	674
Transport and communication	1 125.97	40.9	2 521.15	38.5	224	4 073.80	32.8	362
Trade	0.00	0.0	0.00	0.0	100	0.00	0.0	100
Other	504.21	18.3	1 306.74	19.9	259	2 536.40	20.4	503
Grand total	2 750.36	100.0	6 551.03	100.0	238	12 430.01	100.0	452

a) Components may not add to totals because of rounding.
b) Less than 0.01 million US dollars.
Source: Annex Table A-21.

Table 43. **Soviet exports of capital goods to Eastern Europe, by end-use branch, 1970, 1976 and 1980** [a]

End-use Branch	1970		1976			1980		
	Value (Millions of US dollars)	% of total	Value (Millions of US dollars	% of total	Index (1970 = 100)	Value (millions of US dollars)	% of total	Index (1970 = 100)
Electricity	38.09	3.9	300.24	9.2	788	644.98	12.4	1 693
Fuel and mining	27.05	2.8	214.92	6.6	795	249.41	4.8	922
Metallurgy	13.97	1.4	102.75	3.1	736	275.93	5.3	1 975
Engineering and metalworking	79.31	8.1	259.27	7.9	327	316.13	6.1	399
Chemicals	12.86	1.3	93.61	2.9	728	109.35	2.1	850
Building materials	1.43	0.1	53.16	1.6	3 717	62.40	1.2	4 364
Wood and paper	1.91	0.2	20.08	0.6	1 051	39.13	0.8	2 049
Textile	0.00 (b)	0.0	0.00	0.0	100	0.00	0.0	100
Other light industry	24.04	2.5	77.00	2.3	320	140.68	2.7	585
Food and food processing	12.98	1.3	30.22	0.9	233	39.53	0.8	305
Total industry	211.64	21.6	1 151.25	35.1	544	1 877.55	36.1	887
Construction	63.57	6.5	154.87	4.7	244	277.44	5.3	436
Agriculture, including tractors	165.86	16.9	485.90	14.8	293	764.60	14.7	461
Transport and communication	425.59	43.4	1 124.55	34.3	264	1 735.83	33.3	408
Trade	0.00	0.0	0.00	0.0	100	0.00	0.0	100
Other	114.47	11.7	361.45	11.0	316	550.36	10.6	481
Grand total	981.13	100.0	3 277.72	100.0	334	5 205.78	100.0	531

a) Components may not add to totals because of rounding.
b) Less than 0.01 million US dollars.
Source: Annex Table A-22.

For reasons explained above, one cannot satisfactorily measure the extent to which the increase in Soviet exports to Eastern Europe can be attributed to Soviet imports of Western technology. However, it is possible to identify illustrative instances of this process.

For example, about $2 billion of Western pipe and compressor station equipment constituted critical inputs for the Soiuz Pipeline transporting natural gas from the Soviet Orenburg gasfields to Eastern Europe. For the six East European countries as a group, Orenburg natural gas provided an estimated 17.4 per cent of their total natural gas consumption in 1980, although this percentage share varied by country from 41.8 for Bulgaria to 3.8 for Romania[36].

Soviet production of the Zhiguli-Lada with FIAT technology (see Chapter 4, Section B) involves an extensive scheme of industrial co-operation with East European countries. Under it the latter produce spare parts, components, and sub-assemblies for cars built in the USSR, in return for deliveries of completed vehicles[37]. In turn, the production of diesel trucks at the Kama Plant, constructed with Western technology, enabled the USSR to expand truck exports to East European countries[38].

Western firms supplied complex installations for saw mill and pulp-making operations, supervised construction and assembly, and trained personnel for the "Soviet-East European joint project" to develop the Ust-Ilimsk timber complex[39].

Also, it is possible that some of Soviet exports of numerically-controlled metalworking machine tools, other machinery, mining equipment, and chemicals to Eastern Europe[40] involve technology previously obtained by the USSR from the West. But from the published information one cannot establish the extent of such linkages.

D. SOVIET TRADE WITH LESS DEVELOPED MARKET ECONOMIES

Trade with less developed market economies (called developing market economies in United Nations statistics, on which Tables 44 and 45 are based) accounts for about one-eighth of total Soviet imports and of total Soviet exports, according to Soviet official foreign trade statistics.

The commodity composition of Soviet imports from these countries is shown in Table 44. The dominant commodity group is food, beverages, and tobacco, with almost three-fourths of the total in 1982. With the addition of crude materials and mineral fuels, the share of primary products reaches almost 90 per cent of the total. Soviet imports of machinery and equipment from these countries are negligible, and the potential for technology transfer from them to the USSR is slight.

In the commodity composition of Soviet exports to less developed market economies (see Table 45), the leading specified categories are machinery and transport equipment, and mineral fuels, with respective percentage shares of 20.8 and 20.2 in 1982. However, still larger than either of these is the unspecified category (SITC 9), consisting chiefly of arms and constituting 45.7 per cent of the total in 1982[41].

There is no evidence that Soviet imports of Western technology have significantly affected arms sales by the USSR, which is a major military power with a self-sufficient arms industry and to which the export of technology of military importance is severely restricted by Western governments (see Chapter 2, Section C). Western technology may possibly have contributed in some way to other Soviet exports to less developed market economies – for instance, machine tools and other types of machinery and equipment – but the extent is likely to be small.

Table 44. **Commodity composition of Soviet imports from developing market economies, 1970 and 1976-1982** [a]

Per Cent

Commodity Group	1970	1976	1977	1978	1979	1980	1981	1982
Food, beverages and tobacco (SITC 0 and 1)	48.0	55.7	56.8	63.5	58.4	62.5	66.9	70.7
Crude materials, excluding fuels (SITC 2 and 4)	32.6	16.9	17.9	12.3	13.5	15.2	17.0	13.8
Mineral fuels and related materials (SITC 3)	2.0	13.7	12.8	12.6	16.3	8.8	4.4	4.3
Sub-total, Primary products	82.6	86.5	87.5	88.5	88.1	86.5	88.3	88.8
Chemicals (SITC 5)	1.3	1.6	2.0	1.8	1.4	2.1	1.1	1.1
Machinery and transport equipment (SITC 7)	0.1	0.5	0.4	0.4	0.3	0.5	0.6	0.6
Other manufactured goods (SITC 6 and 8)	15.9	11.4	10.1	9.3	9.6	10.3	9.6	9.0
Sub-total, manufactures	17.3	13.5	12.5	11.5	11.4	12.9	11.3	10.6
Commodities and transactions not classified according to kind (SITC 9)	0.1	0.0	0.0	0.0	0.5	0.5	0.5	0.5
Total	100.0	100.0	100.0	100.0	100.0	100.0	100.0	100.0

a) Components may not add to totals because of rounding.
Source: Computed from United Nations trade statistics in millions of US dollars in the following sources: 1970, 1976-77: United Nations, *Yearbook of International Trade Statistics, 1981* (New York, 1982), Vol. 1, Special Table C. 1978-1979: United Nations, *Monthly Bulletin of Statistics*, Vol. 37, No. 5 (May 1983), Special Table E. 1980-1982: United Nations, *Monthly Bulletin of Statistics*, Vol. 38, No. 5 (May 1984), Special Table D.

Table 45. **Commodity composition of Soviet exports to developing market economies, 1970 and 1976-1982** [a]

Per Cent

Commodity Group	1970	1976	1977	1978	1979	1980	1981	1982
Food, beverages and tobacco (SITC 0 and 1)	6.8	4.8	4.1	4.1	4.4	4.0	3.9	3.7
Crude materials, excluding fuels (SITC 2 and 4)	5.1	4.7	3.8	3.3	2.9	3.4	5.3	3.6
Mineral fuels and related materials (SITC 3)	5.6	12.8	12.1	11.5	16.9	18.4	20.0	20.2
Sub-total, Primary products	17.5	22.3	20.0	19.0	24.2	25.8	29.2	27.4
Chemicals (SITC 5)	1.8	2.0	1.8	1.9	1.6	2.1	3.0	2.4
Machinery and transport equipment (SITC 7)	33.7	25.7	21.9	23.6	23.8	24.1	21.1	20.8
Other manufactured goods (SITC 6 and 8)	10.2	5.8	5.2	5.2	4.6	4.2	3.6	3.6
Sub-total, Manufactures	45.7	33.5	28.9	30.8	30.0	30.4	27.6	26.8
Commodities and transactions not classified according to kind (SITC 9)	36.8	44.2	51.1	50.2	45.8	43.8	43.2	45.7
Total	100.0	100.0	100.0	100.0	100.0	100.0	100.0	100.0

a) Components may not add to totals because of rounding.
Source: Computed from United Nations trade statistics in millions of US dollars in the following sources: 1970, 1976-77: United Nations, *Yearbook of International Trade Statistics, 1981* (New York, 1982), Vol. 1, Special Table C. 1978-1979: United Nations, *Monthly Bulletin of Statistics*, Vol. 37, No. 5 (May 1983), Special Table E. 1980-1982: United Nations, *Monthly Bulletin of Statistics*, Vol. 38, No. 5 (May 1984), Special Table D.

Some Soviet exports to less developed market economies occur through "tripartite" industrial co-operation (TIC), involving the participation of Western and Soviet firms in building manufacturing and power generation plants in these countries. The available information on Soviet TIC activities is scanty. The pertinent statistical data often refer to numbers of projects (including those under discussion as well as those actually undertaken), rather than the value, nature, or technological level of the Soviet and the Western participation. TIC is not an important factor in Soviet capital goods exports to these countries. Nor is there evidence that the USSR gains through TIC Western technology otherwise unavailable to it[42].

Thus, there is no significant indication that developing market economies transfer to the USSR technology they have acquired from the West, or that the USSR sells to the developing market economies technology it previously obtained from the West.

NOTES AND REFERENCES

1. "Soviet Foreign Trade Performance in 1983", Wharton Econometric Forecasting Associates (WEFA), *Centrally Planned Economies Current Analysis,* 9th April 1984.
2. *Ibid.*
3. "Soviet Foreign Trade, January-December 1983 (Statistical Data)", *Foreign Trade* (Moscow), 1984, No. 3, supplement.
4. *Ibid.*
5. *Ibid.*
6. For a more detailed explanation of these problems, see Jan Vanous, "Soviet and East European Foreign Trade in the 1970s: A Quantitative Assessment", in US Congress, Joint Economic Committee, *East European Economic Assessment, Part 2 – Regional Assessments,* 97th Congress, 1st Session (Washington, D.C.: US Government Printing Office, 1981), pp. 687-688 and 712-714. For an effort to calculate Soviet foreign trade with Eastern Europe at world market prices, see Michael Marrese and Jan Vanous, *Soviet Subsidization of Trade with Eastern Europe: A Soviet Perspective* (Research Series, No. 52; Berkeley, Calif.: University of California Institute of International Studies, 1983).
7. On discrepancies in "mirror" statistics, see Alexander J. Yeats, "On the Accuracy of Partner Country Trade Statistics", *Oxford Bulletin of Economics and Statistics,* Vol. 40, No. 4 (November 1978), pp. 341-361. The classification schemes of Western and Eastern foreign trade statistics are compared in United Nations, Statistical Commission and Economic Commission for Europe, *Correspondence Table Between the Standard International Trade Classification of the United Nations (SITC) and the Standard Foreign Trade Classification of the Council for Mutual Economic Assistance (SFTC)* (Conference of European Statisticians, Statistical Standards and Studies, No. 32; New York, 1982). For a detailed discussion of the issues in the present context, see Damian T. Gullo, "Reconciliation of Soviet and Western Trade Data: The United States as a Case Study", in US Congress, Joint Economic Committee, *Soviet Economy in a Time of Change,* Vol. 2, pp. 526-550.
8. See Drabek and Slater, *Methodology.*
9. For further discussion of the R & D intensity of OECD countries' exports to the USSR (and also to Eastern Europe), see Zdenek Drabek, "The Impact of Technological Differences on East-West Trade", *Weltwirtschaftliches Archiv,* Vol. 119 (1983), No. 4, pp. 630-648.
10. See, respectively, Zdenek Drabek, *Western Embodied Technology and its Sectoral Impact on East European Exports to the West,* working document, OECD, Paris, 1981, pp. 41-45, and Philip Hanson, "The End of Import-Led Growth? Some Observations on Soviet, Polish, and Hungarian Experience in the 1970s", *Journal of Comparative Economics,* Vol. 6, No. 2 (June 1982), pp. 130-147. For Drabek's critique of Hanson's work, see Drabek, *Western Embodied Technology,* pp. 45-47.
11. Cf. OECD, *Recent Developments in Countertrade,* p. 31.

12. Savin, "Co-operation", p. 24.
13. S. Ponomaryov, "Compensation-Based Co-operation between the USSR and the Capitalist Countries in the Fuel Industry", *Foreign Trade* (Moscow), 1978, No. 4, p. 28.
14. Hewett, "Near-Term Prospects", pp. 409-411; Hannigan and McMillan, *The Soviet-West European Energy Relationship, pp. 59-62;* and Jonathan P. Stern, "Specters and Pipe Dreams", *Foreign Policy,* No. 48 (Fall 1982), p. 23.
15. All such estimates involve assumptions about, for example, rates of economic growth in West European countries, changes in oil prices and the relationship of gas prices to oil prices, and the availability and terms of gas deliveries from sources other than the USSR.
16. On trends in Soviet exports of chemicals to the West, see OECD, *East-West Trade in Chemicals,* pp. 32-35; Vadim Borin, "The Export Potential of the Soviet Chemical Industry", *Foreign Trade* (Moscow), 1980, No. 8, pp. 10-11 and 22; and V. Molodtsov, "Dynamic Trade in Chemical Products", *Foreign Trade* (Moscow), 1981, No. 10, pp. 15-19.
17. Under strict definitions, the large Soviet agreement with Occidental Petroleum is a barter rather than a compensation deal. Over the period 1978-1997, it provides for Soviet imports of $10 billion in superphosphoric acid and technology and equipment for fertilizer production, and Soviet exports of $10 billion worth of ammonia, urea, and potash. For details of this agreement, see OECD, *East-West Trade in Chemicals,* p. 49 and Annex 1, p. iii.
18. OECD, *East-West Trade in Chemicals,* pp. 40-41.
19. *Ibid.,* p. 49.
20. *Ibid.,* pp. 45-60.
21. In 1983 the Commission of the European Communities imposed an antidumping duty on imports into the European Economic Community of hexamethylenetetramine from the Soviet Union (and the German Democratic Republic). See Commission Regulation (EEC) No. 348/83 of 10th February 1983, *Official Journal of the European Communities,* No. L 40 (12th February 1983), pp. 24-27. In response to a petition from US ammonia producers, claiming Soviet dumping in the US market, the US International Trade Commission determined that there was injury to US producers and set an import quota for Soviet ammonia, but this decision was later overturned. See Hedija H. Kravalis, "USSR: An Assessment of US and Western Trade Potential with the Soviet Union through 1985", in US Congress, Joint Economic Committee, *East-West Trade: The Prospects to 1985,* 97th Congress, 2d Session (Washington, D.C.: US Government Printing Office, 1982), p. 289.
22. Valery Popyrin, "Compensation Agreements – An Effective Form of Mutually Profitable Trade", *Foreign Trade* (Moscow), 1980, No. 3, p. 22.
23. USSR, Ministerstvo vneshnei torgovli, Glavnoe planovo-ekonomicheskoe upravlenie, *Vneshniaia torgovlia SSSR v 1982 g.: Statisticheskii sbornik* [Foreign trade of the USSR in 1982: Statistical Handbook] (Moscow: Finansy i statistika, 1983), pp. 60-61.
24. Gutman, "East-West Industrial Co-operation", pp. 20-24, and Kravalis, "USSR: An Assessment", pp. 291-292.
25. Dmitri Petrov and Yuri Savinov, "Structural Shifts on World Equipment Markets and Ways of Soviet Machinery Export Development", Part 2, *Foreign Trade* (Moscow), 1982, No. 3, pp. 26-27.
26. Only about 4 per cent of total Soviet machinery exports go to developed market economies, because "...a significant quantity of [Soviet] machinery and equipment is insufficiently competitive not only on the capitalist market but also in socialist countries, where our goods collide with others furnished by capitalist countries", according to Smeliakov, "I spros".
27. Such competition is sometimes called a "boomerang effect" – or a "ricochet effect" if the competition is in a third market rather than the home market of the Western country from which the technology was sold to the USSR. Cf. Gutman, "East-West Industrial Co-operation", p. 20.

28. On such clauses, see Gutman, "East-West Industrial Co-operation", p. 24, and Holliday, *Transfer of Technology*, p. 19.

29. Organisational arrangements for the transfer of disembodied technology among the USSR and the East European countries are discussed in Louvan E. Nolting, *Integration of Science and Technology in CEMA* (Foreign Economic Report No. 21; Washington, D.C.: US Department of Commerce, Bureau of the Census, 1983).

30. Alexei Stromov, "The Tenth Five-Year Plan Period: Soviet Machinery and Equipment Imports from the CMEA Countries", *Foreign Trade* (Moscow), 1981, No. 6, p. 27.

31. "Soviet Foreign Trade Performance".

32. Stromov, "The Tenth Five-Year Plan Period", p. 22.

33. Michael R. Dohan, "Export Specialisation and Import Dependence in the Soviet Economy, 1970-1977", in US Congress, Joint Economic Committee, *Soviet Economy in a Time of Change*, Vol. 2, pp. 355-361.

34. Philip Hanson, "Soviet Trade with Eastern Europe", in Karen Dawisha and Philip Hanson (eds.), *Soviet-East European Dilemmas: Coercion, Competition, and Consent* (London: Heinemann, 1981), p. 103.

35. Kenneth Tasky, "Eastern Europe: Trends in Imports of Western Computer Equipment and Technology", in US Congress, Joint Economic Committee, *"East European Economic Assessment, Part 2 – Regional Assessments"*, pp. 319-327.

36. J.B. Hannigan, *The Orenburg Natural Gas Project and Fuels-Energy Balances in Eastern Europe* (East-West Commercial Relations Series, Research Report No. 13; Ottawa, Canada: Carleton University Institute of Soviet and East European Studies, 1980), pp. 40 and 54.

37. Gutman, "East-West Industrial Co-operation", pp. 43-45, provides a detailed account.

38. I. Schweitzer, "Some Particularities of Hungarian Machine Imports from the Soviet Union", *Acta Oeconomica* (Budapest), Vol. 18 (1977), No. 3-4, p. 340.

39. See Chapter 4, Section B, Sub-section 5.

40. Nikolai Ivanov, "USSR and the CMEA Member-Countries' Co-operation in the Planned Provision of the Means of Production", *Foreign Trade* (Moscow), 1982, No. 4, pp. 8-11, and Ivan Kapranov, "With the Assistance of the Soviet Union", *Foreign Trade* (Moscow), 1983, No. 6, pp. 4-5.

41. Data on Soviet arms sales are not published by the USSR or most of the recipient LDCs. Western estimates of the magnitude and composition of arms sales vary. For a discussion of different estimates of the value of these sales, see Vanous, "Soviet and East European Foreign Trade", pp. 692-693. On the numbers of each type of Soviet arms exports (e.g. tanks, naval vessels, aircraft, missiles, etc.) and their geographical distribution, see Richard F. Grimmett, *Trends in Conventional Arms Transfers to the Third World by Major Supplier, 1976-1983* (Report No. 84-82 F; Washington, D.C.: Congressional Research Service, 1984). On the reasons for Soviet arms deliveries to these countries, see, for example, Roger E. Kanet, "Soviet and East European Arms Transfers to the Third World: Strategic, Political, and Economic Factors", in *External Economic Relations of CMEA Countries: Their Significance and Impact in a Global Perspective* (Brussels: NATO, 1984), pp. 171-194.

42. On TIC, see Carl H. McMillan, *The Political Economy of Tripartite (East-West-South) Industrial Co-operation* (East-West Commercial Relations Series, Research Report No. 12; Ottawa, Canada: Carleton University Institute of Soviet and East European Studies, 1980), and Patrick Gutman, "Tripartite Industrial Co-operation and East Europe", in US Congress, Joint Economic Committee, *East European Economic Assessment, Part 2 – Regional Assessments*, pp. 823-871.

Chapter 6

CONCLUSION

Technology is knowledge necessary to apply a process, manufacture a product, or provide a service. This knowledge includes:

1. Technical information about process or product characteristics;
2. Production techniques for the transformation of labour, materials, components, and other inputs into finished outputs; and
3. Managerial systems to select, schedule, control, and market production.

Technology may be "disembodied", for instance in technical documentation, or "embodied", for example in machinery and equipment.

"Material transfer" of technology occurs with the import of a product by a country that cannot produce it. In the case of "design transfer" the receiver imports a plant to make the product. "Capacity transfer" entails the receiver's ability to replicate the imported plant by building additional similar facilities without foreign help. The broad conception of technology transfer includes "material transfer" because the importing country, by using the foreign machinery, equipment, or intermediate product, thereby obtains the benefits of a technology not otherwise available to it. This comprehensive approach helps to explain important facets of Soviet technological development and Soviet economic relations with Western countries.

The USSR has imported Western technology in many fields in an effort to stimulate technological progress in order to promote the growth and modernisation of the economy. In the last two decades most of Soviet economic growth has been due to increased inputs of factors, especially capital, rather than to improvements in factor productivity, which has grown slowly or decreased. As increments to the labour force and the capital stock declined, the Soviet regime shifted to an "intensive" growth strategy emphasising increases in factor productivity. Gains in factor productivity can come from many sources. They include not only technological progress in the strict sense of the introduction of new processes and products, but also economies of scale, reallocation of factors across industries, better organisation and management, higher quality of labour inputs, and greater effort.

Western statistical analyses suggest that lagging technological progress has depressed the rates of growth of factor productivity and output in the USSR. This view is supported by studies of systemic handicaps to Soviet technological progress and by assessments of the technological lag of Soviet industry behind the West.

Systemic handicaps include problems in organisation, financing, supply, pricing, and performance indicators and incentives characteristic of a "tautly planned" centrally administered economy.

Because there are various methods of measuring comparative technological levels, it is not possible to reach simple summary conclusions about the Soviet technological lag. But detailed studies have identified significant lags in the iron and steel, chemical, energy, machine tool, motor vehicle, and computer industries, for instance.

The Soviet Union's effort to acquire Western technology for these and other branches of the economy has been constrained by three principal factors: Soviet ability to assimilate Western technology, Soviet capacity to finance technology imports for convertible currency, and Western governments' restrictions on the export of technology to the USSR.

Because Western technology is both foreign and more advanced than Soviet technology, its introduction is likely to be more difficult than the introduction of Soviet domestic technology.

The USSR has a deficit in its convertible currency trade that is covered chiefly by sales of arms and gold and by borrowing. Also, the USSR has sought to finance imports through countertrade, primarily compensation agreements. These agreements link the debt service payments on loans for imports of Western machinery and equipment to a contracted flow of Soviet convertible currency exports from projects incorporating the imports.

Western governments have imposed multilateral or unilateral restrictions on exports to the USSR of technology deemed to have military significance or important effects on Soviet economic development and exports. Such restrictions have affected, for example, exports of Western technology in the electronics, electrical machinery, metalworking, oil and gas, and chemical industries.

Despite these several kinds of constraints, the USSR has imported a significant amount and range of Western technology through a variety of modes of transfer.

One mode is licencing. Although relevant statistical data are scanty, the USSR has shown a preference for buying "associated" licences linked with purchases of machinery and equipment from the West, and concentrated in the chemical, electrical equipment, iron and steel, and machine tool industries.

Commodity imports can be regarded as a mode of technology transfer when they embody technologies not available in the recipient country. Such imports transfer not the methods of production themselves, but rather the products constituting the results of these technologies. In some cases, a nation which possesses the relevant technology, but has not diffused it widely enough, may import part of its supply in order to supplement domestic production. However, not all imports can be explained by technological lags and embodied technology transfers. The level and composition of imports (and exports) also reflect inter-industry and intra-industry specialisation decisions made for comparative advantage and other reasons.

Soviet imports of manufactured goods from the OECD countries have been concentrated in machinery, equipment, metal products, and chemicals, and they have been destined primarily for the Soviet engineering and metalworking, chemical, and energy industries. The share of "technology-based" products in total exports of OECD countries to the USSR was 57 per cent in 1982. The other 43 per cent of OECD countries' exports to the USSR consisted of products judged not to be "technology-based". The combined share of goods of "high" and "moderate" R & D intensity in OECD countries' total exports to the USSR ranged from 7 to 14 per cent during 1970-1982.

Turnkey projects are a much more comprehensive and potentially more effective mode of technology transfer than separate commodity imports or licence purchases. In turnkey projects, foreign firms supply whole production systems – designing facilities, supervising construction and installation work, and training personnel. Turnkey projects have been used in the Soviet chemical, ferrous metallurgy, motor vehicle, machine building, and light and food industries, for instance.

Industrial co-operation agreements (ICAs) provide continuing involvement of a Western firm in a Soviet firm's production methods, product characteristics, and quality control. Soviet ICAs with Western firms have been concentrated in the chemical, machine building, electrical equipment, and transport equipment industries.

The USSR also obtains Western technology through a variety of "non-commercial" channels, including publications and trade shows, scientific exchange programmes, and illegal methods.

It is difficult to assess the impact of imported Western technology on the Soviet economy, for a number of reasons. First, appropriate specific quantitative data are available for commodity imports but not for licences and ICAs. Second, it is hard to measure the effects even of imports of Western machinery and equipment on the growth of Soviet capital stock, output, and exports. Third, because imports of Western technology are small in comparison with the size of the Soviet economy, these effects are likely to be small for the economy as a whole, but they could be significant for particular branches, sub-branches, or product groups. Fourth, one ought to consider not only the direct impact of technology transfer on the output of a recipient Soviet enterprise, but also various indirect impacts, for instance on supplier and customer enterprises.

Macroeconomic estimates of the impact of Western technology on the Soviet economy are controversial because they rely on simplifying assumptions of disputed validity as well as on incomplete and imperfect statistical data. The dominant view among specialists is that the aggregate impact of the transfer of Western technology on Soviet industry cannot really be measured satisfactorily but is likely to be rather modest (and the impact on the Soviet economy as a whole, of which industry is only a part, even more so).

Nonetheless, significant effects from the transfer of Western technology can be identified in a number of Soviet industries, including chemicals, motor vehicles, machine tools, energy, and forest products.

However, the impact of Western technology depends upon the speed and extent of its assimilation by the Soviet economy. Although Western knowledge of Soviet assimilation experience is limited, the USSR appears to encounter problems in both phases of the assimilation process. The initial absorption phase involves exploitation of Western technology in the first facility for which it is acquired. Problems in this phase include delays between the delivery of Western machinery and equipment and their installation and use in production, and lower performance (once in production) than in the West. Both problems are due to shortages of complementary inputs of Soviet origin. The subsequent diffusion phase entails the replication of Western technology in other plants. It also is hampered by shortages of labour, materials, and equipment, as well as by problems in copying or adapting foreign technology.

Manufactured goods (especially engineering products and iron and steel products) dominate Soviet imports from developed capitalist market economies. However, the share of food products has been significant in recent years as a result of heavy Soviet grain purchases in response to poor Soviet harvests. In contrast, primary products represent about four-fifths – and fuels alone more than three-fourths – of Soviet exports to developed market economies. "Technology-based" goods account for a little more than half of OECD countries' exports to the USSR but only one-tenth of Soviet exports to OECD countries.

For some product groups there are strong links between Soviet imports of technology from the West and subsequent Soviet exports to the West. In some cases the two flows are explicitly connected under compensation agreements – for instance, agreements providing for Soviet exports of chemicals from facilities built with Western plant and equipment and of natural gas through pipelines constructed with Western large-diameter pipe and compressor station equipment. A clear link exists also for Soviet exports of passenger cars based on Western technology.

The transfer of Western technology to the USSR has also affected Soviet trade with Eastern Europe. On the one hand, the USSR has been able to obtain from the West machinery and equipment technologically superior to East European products. On the other, Soviet imports of Western technology have played a critical role in important CMEA "integration projects", like the Soiuz natural gas pipeline, the motor vehicle programme, and exploitation of forest resources.

In contrast, the import of Western technology appears to have little influence on Soviet trade with non-Communist less developed countries.

The USSR will continue to import Western technology through a variety of transfer modes for many branches of the Soviet economy. Average annual growth rates in the USSR in 1984-1988 are likely to be less than 0.1 per cent for the labour force and in the 3-4 per cent range for investment[1]. Hence, the Soviet regime will strive to achieve improvements in factor productivity, in which technological progress plays a key role. However, the systemic handicaps to technological progress in the USSR are serious. There is no evidence of significant success in overcoming them through the various measures taken to improve the research, development, and innovation process[2]. Nor is technological progress – or economic growth generally – likely to be increased much by Soviet efforts at economic system reforms undertaken since 1979[3]. Thus the Soviet need for Western technology will remain strong.

But the scope and content of future Soviet acquisition of Western technology will depend upon a number of factors. They include the nature of the Soviet investment programme, the assessment of the USSR's experience in the acquisition and assimilation of Western technology, financing constraints, and Western governments' export controls.

Econometric analysis of the "determinants" of Soviet trade with the West shows that Soviet machinery imports from the West are strongly correlated with Soviet investment in machinery, although relative prices of Western and East European machinery have some influence[4]. The investment programme in the 1981-1985 Soviet economic plan reiterates the emphasis of earlier five-year plans on renovation and re-equipment of existing facilities, in preference to the construction of new facilities. Thus, a larger share of investment should be devoted to machinery and equipment and a smaller share to buildings. For example, 85 per cent of the increase in machinery output during 1981-1985 is to come from renovation. Soviet investment specialists consider renovation and re-equipment superior to building new facilities. The former involve chiefly machinery and equipment, from one of the relatively more efficient branches of the Soviet economy, and less construction, which is characterised by long delays and large cost overruns. Also, renovation and reconstruction hasten the withdrawal of old technology and its replacement with new, resource-saving technology[5].

However, this renovation and re-equipment strategy has not been successfully implemented. First, the replacement of machinery and equipment often requires considerable construction work for remodelling, re-engineering, and even expansion of facilities. Second, project design units prefer to concentrate on design work for new enterprises, which is easier and more profitable than reconstruction projects. In turn, machine building enterprises prefer to make standardized equipment for new factories, rather than machines to fit the specific conditions of existing facilities. Third, both the producing enterprise and the construction organisation like a new facility better than the renovation or re-equipment of an old one. Renovation interferes with production activity and the achievement of plan targets and associated bonuses. Construction enterprises find it easier to work on open building sites and thus to report a larger volume of work for plan fulfilment[6].

The Soviet programme for fixed investment in 1981-1985 was planned at a modest 10 per cent above the investment accomplished during 1976-1980. However, within the 1981-1985 total, branch priorities were sharply altered. Investment in fuel and power is planned at 52 per cent, and investment in ferrous metallurgy at 30 per cent, above the respective 1976-

1980 amounts. In contrast, investment in machine building and metalworking during 1981-1985 is scheduled to be only 10 per cent above the 1976-1980 level. Within fuel and power, in 1981-1985 investment in natural gas production is to grow 120 per cent, and in oil production 63 per cent, above the 1976-1980 amounts. In addition, investment in pipeline construction (included in the transportation sector) is scheduled to be 45 per cent higher in 1981-1985 than in the preceding five years[7].

The emphasis in the current investment programme on fuel production and transport stems from the USSR's urgent need to expand energy output for domestic consumption, for East European requirements, and for convertible currency exports to the world market[8]. The emphasis of the Soviet investment programme on energy also has been responsible for changes since the mid-1970s in the composition of Soviet imports of capital goods from OECD countries. The share of pipe, compressor station equipment, and mining machinery has risen, while the share of machinery and equipment for the machine-building and chemical industries has declined.

Although the investment programme of the next five-year plan, for 1986-1990, has not yet been released, it is likely to continue the emphasis on the fuel and raw material branches, in view of the USSR's domestic and export requirements for their output. The extent to which the investment programme involves the acquisition of Western technology will depend upon the Soviet assessment of its past technology transfer experience, the Soviet ability to pay for technology from developed market economies, and Western governments' export controls.

The USSR may buy Western machinery and equipment because of their technological superiority or, when the USSR (or an East European country) has a suitable technology, because of inadequate production capacity in the USSR (or Eastern Europe). There is disagreement among the relevant decision-makers and their specialist advisors about the extent to which Western technology should be chosen in preference to Soviet (or East European) technology[9].

One school argues that there are many instances in which Western technology is clearly superior in all pertinent characteristics (such as productivity, weight, reliability, service life, etc.). In such cases, according to this view, Western machinery or equipment, or licences covering their technology, should be purchased, in order to get earlier and better production of the desired output in the USSR. In these cases, one should override the opposition to such technology imports voiced by R & D organisations that have tried unsuccessfully to develop a corresponding Soviet technology[10].

Various arguments are presented to curtail the import of Western technology. Assimilation problems have been discussed above. Also, it is asserted that branch ministries and foreign trade organisations sometimes seek to, and do, buy foreign equipment or licences when available Soviet technology is equal or better[11]. Even if Western technology is superior, it has the disadvantage of requiring expenditure of convertible currency, not only initially for the purchase of the machinery or equipment but also subsequently for spare parts and for associated materials (like lubricants) not available in the USSR[12]. Further, there is a risk that Western deliveries of machinery, spare parts, and materials might be reduced or stopped by export restrictions of one or more Western governments[13].

Financing constraints also may make Soviet acquisition of Western technology more modest in the future than in the last decade. The Soviet ability to buy Western embodied or disembodied technology depends primarily on convertible currency earnings from oil and gas, gold sales, and arms sales; outlays of convertible currency for grain; and credit from OECD countries. It is possible to identify some developments in these factors that could enhance or diminish the USSR's convertible currency resources for the purchase of Western technology. But the number of variables is so large, and the uncertainty about them so great, that it would be far beyond the scope of this study to attempt any projections or forecasts of the future outcome[14].

Soviet ability to finance future technology imports from the West will benefit insofar as one or more of the following occurs:

1. The volume of Soviet oil sales for convertible currency is sustained, through some combination of:
   a) Stabilization or only a small reduction in output;
   b) Domestic conservation;
   c) Curtailment of oil deliveries to Eastern Europe; and
   d) Re-exports of oil from Middle Eastern countries.
2. World market oil prices do not fall, or they rise.
3. The quantities and prices of Soviet natural gas deliveries to Western Europe are larger.
4. World market prices for gold are higher.
5. The USSR has greater opportunities for arms sales to Third World countries for payment in convertible currency (or oil).
6. Soviet grain harvests are larger, as a result of good weather and/or the progress of the "Food Programme"[15].
7. World market grain prices are lower.
8. Western credit is available to the USSR on more favourable terms, for instance because weaker business conditions in developed market economies lead to lower interest rates.

Conversely, Soviet ability to finance technology imports from the West will be weakened by lower quantities and/or prices for convertible currency exports of oil, gas, gold, and arms; by higher quantities and/or prices for convertible currency imports of grain; and/or by stricter terms for private and official Western credit.

A tightening of unilateral or multilateral export controls by Western governments will affect the composition and perhaps the total amount of Soviet technology imports from developed market economies. The USSR has responded to such a curtailment of Western exports with efforts to produce substitutes in the Soviet Union or elsewhere in the CMEA area. According to Nikolai Tikhonov, Chairman of the USSR Council of Ministers:

> ...In the past few years, the Soviet Union has organised the production of many articles of the chemical, metallurgical, and machine-building industries that in the past had been bought from the United States or countries that support the American leadership's discriminatory actions. Additional assignments in this field have been established for 1984-1990. It would be useful to prepare concrete proposals on the implementation by the CMEA countries of measures to organise the joint production of several machines, types of equipment, and materials on the sale of which the West has imposed restrictions[16].

Oleg Bogomolov, Director of the USSR Academy of Sciences' Institute of the Economics of the World Socialist System, has suggested possible changes in technology acquisition policies in response to Western export restrictions:

> ...The strategy of the socialist countries regarding importation of Western technology will, it seems, be directed away from the purchase of individual items of equipment for specific branches of industry, means of transportation, and pipes, and towards the purchase of the complete equipment and technology for domestic manufacture of the corresponding types of output. In the future we will continue to show interest in the

purchase of licences, but mainly in those cases where we possess our own research facilities to improve further the technology that has been purchased[17].

The Soviet leadership thus faces various policy issues affecting the acquisition of Western technology:

1. What will be the size and composition of the investment programme?
2. To what degree must Western technology surpass Soviet (or East European) technology in order to justify the choice of Western technology, despite its convertible currency cost and its vulnerability to possible Western export restrictions?
3. To what extent should technology acquisition policy stress "design transfer" rather than "material transfer"?
4. How much convertible currency should the USSR try to mobilise for the acquisition of Western technology – for instance, by redirecting oil exports from Eastern Europe to Western Europe, drawing on gold reserves, restraining grain purchases, and/or increasing the convertible currency debt-service ratio?

In turn, there are policy issues for Western governments concerning the transfer of technology to the USSR:

1. To what degree and in what ways should unilateral or multilateral export restrictions of Western governments be increased or reduced?
2. What will be the role for Soviet oil and gas in the energy supply of Western Europe?
3. In what amounts and on what terms should Western countries provide official or private credit to the Soviet Union?

In view of the Soviet need for Western technology, on the one hand, and the constraints on its acquisition, on the other, it is reasonable to expect that Soviet imports of Western technology will continue, but on a more modest scale and a more selective basis than in the past decade. Proposed acquisitions of Western technology will be evaluated according to stricter standards for the priority of different industries (branches, product groups), the superiority of Western technology over Soviet (or East European) alternatives, the probable ease of assimilation of Western technology, and its potential contribution to increasing convertible currency exports or reducing convertible currency imports.

NOTES AND REFERENCES

1. Wharton Econometric Forecasting Associates, *Centrally Planned Economies Outlook* (Washington, D.C., September 1983), p. 57. For a concise discussion of major current problems retarding Soviet economic growth, see Boris Rumer, "Structural Imbalance in the Soviet Economy", *Problems of Communism,* Vol. 33, No. 4 (July-August 1984), pp. 24-32.
2. Ronald Amann, "Industrial Innovation in the Soviet Union: Methodological Perspectives and Conclusions", in Amann and Cooper, *Industrial Innovation,* pp. 30-37.
3. See Morris Bornstein, "Improving the Soviet Economic Mechanism", *Soviet Studies,* Vol. 36, No. 1 (January 1985), and Fyodor I. Kushnirsky, "The Limits of Soviet Economic Reform", *Problems of Communism,* Vol. 33, No. 4 (July-August 1984), pp. 33-43.

4. "The Determinants of East-West Trade", United Nations, Economic Commission for Europe, *Economic Bulletin for Europe,* Vol. 33, No. 4 (December 1981), p. 579.
5. Robert Leggett, "Soviet Investment in the Eleventh Five-Year Plan", in US Congress, Joint Economic Committee, *Soviet Economy in the 1980s,* Part 1, p. 135.
6. On the failure of the renovation and re-equipment strategy, see Boris Rumer, "Soviet Investment Policy: Unresolved Problems", *Problems of Communism,* Vol. 31, No. 5 (September-October 1982), pp. 53-68, and Boris Rumer, "Some Investment Patterns Engendered by the Renovation of Soviet Industry", *Soviet Studies,* Vol. 36, No. 2 (April 1984), pp. 257-266.
7. Leggett, "Soviet Investment", pp. 136-138.
8. See Thane Gustafson, "Soviet Energy Policy", and Laurie Kurtzweg and Albina Tretyakova, "Soviet Energy Consumption: Structure and Future Prospects", respectively pp. 431-456 and pp. 355-390 in US Congress, Joint Economic Committee, *Soviet Economy in the 1980s,* Part 1.
9. For brief reviews of the controversy, see Parrott, *Politics,* pp. 265-278, and Philip Hanson, "Changes in Soviet Policy on Technology Imports", *Osteuropa Wirtschaft,* Vol. 27, No. 2 (June 1982), pp. 128-133.
10. See, for instance, Smeliakov, "I spros", and Anatolii Zlobin and Nikolai Smeliakov, "Pokupaiut luchshee!" [They buy the better one!], *Literaturnaia gazeta,* 2nd May 1984, p. 11.
11. According to L.I. Brezhnev's report to the 26th Communist Party Congress in 1981 (*Pravda,* 24th February 1981, p. 5): "We must look into the reasons that we sometimes lose our priority and spend large sums of money to purchase from foreign countries equipment and technologies that we are fully capable of producing ourselves, and often of a higher quality too". For some examples, see A. Batygin, "Kopeika million berezhet" [A kopeck saves a million], *Pravda,* 14th June 1980, p. 3; Iu. Kuz'ko, "Kupit' proshche?" [Is it easier to buy?], *Pravda,* 2nd June 1981, p. 3; Vasilii Parfenov, "Inzhenery: sotsial'no-ekonomicheskoe obozrenie" [Engineers: a social-economic review], *Pravda,* 29th June 1981, p. 2; and B. Konovalov, "Tsel' nauk–predvidenie i pol'za" [The purpose of the sciences–foresight and benefit], *Izvestiia,* 21st May 1984, p. 3 (an interview with N.S. Yenikolopov, Director of the USSR Academy of Sciences' Institute of Synthetic Polymer Materials).
12. Vasilii Kochek, "Soviet Foreign Trade on the Threshold of the Eleventh Five-Year Plan", *Foreign Trade* (Moscow), 1981, No. 5, p. 14.
13. "Vstupitel'noe slovo prezidenta Akademiia nauk SSSR akademika A.P. Aleksandrova" [Speech of Academician A.P. Aleksandrov, President of the USSR Academy of Sciences], *Vestnik Akademii nauk SSSR,* 1982, No. 6, pp. 10-11.
14. For an examination of the interaction of some of these factors in the recent past, see Grossman and Solberg, *The Soviet Union's Hard-Currency Balance of Payments.*
15. For a sceptical appraisal, see Anton F. Malish, "The Food Programme: A New Policy or More Rhetoric?" in US Congress, Joint Economic Committee, *Soviet Economy in the 1980s,* Part 2, pp. 41-59.
16. "Novye etapi sotsialisticheskoi ekonomicheskoi integratsii-XXXVII zasedanie sessii SEV" [New stages of socialist economic integration–the 37th session of the CMEA], Pravda, 19th October 1983, p. 4.
17. See his interview with P. Negoitsa, "Real'nyi podkhod i politicheskii avantiurizm" [The real approach and political adventurism], *Trud,* 3rd July 1982, p. 3.

ANNEX

Table A-1. **Examples of Soviet compensation agreements with the West**

Western Country	Western Supplier	Year Signed	Type of Soviet Import	Value of Soviet Imports (Million US$)	Type of Soviet Export	First Year of Soviet Deliveries	Value of Soviet Exports (Million US$)		Remarks
					Forestry		1975-80	1981-85	
Japan	KS industries	1969	Forestry handling equipment	163	Timber products	1969-74	n.a.	n.a.	
Japan	Japan Chip Trading Co	1971	Wood chip plant	45	Wood chips and	1972-81	145	50	
Japan	KS industries	1974	Forestry handling equipment	(525) 500-550	Timber products	1975-79	1 100	n.a.	
France	Parsons Withmore	1973	Pulp-paper complex	60	Wood pulp	1977	34	50	Finland and Sweden also participated in transaction.
Japan	Mitsubishi Heavy Industry Ltd. – Mitsubishi Corp.	1976	Plant for making ground pulp	5.5	Ground pulp	1978	n.a.	n.a.	USSR to deliver undisclosed percentage production capacity.
Finland	Enso-Gutzeit	1976	55 000 ton/yr. double cardboard and special boards	n.a.	1.5 million cu.m. of timber and hardwood per year	1976	n.a.	n.a.	1976-80.
					Chemicals				
Germany	Salzgitter	1972	Polyethylene plant capacity: 120 000 t/yr.	63	Polyethylene	1972	170 (1971-1983)		USSR contracted to export between 150 000 and 250 000 tons/yr. of polyethylene to Germany.
Germany	Salzgitter	1973	Polyethylene plant capacity: 240 000 t/yr.	87	Polyethylene	n.a.	225 (1978-1986)		250/350 000 tons of low density polyethylene in total.
France	Litwin S.A.	1973	Styrene/Polystyrene	120	Polystyrene	1978	50	60	
Italy	Montedison	1973 (preliminary agreement)	11 Chemicals plants	(500)*	Ammonia	1978	287 (1978-1987) (ammonia) 78 (1976-1985) (urea)		5 plants ordered so far; contracts signed for one urea and one acrylonitrile plant in 1974; two freon plants in 1975: one urea plant in 1976. * Figure is based on the value of the 7 plants ordred so far; the total value of all 11 plants is given as $800 million.
United	John Brown	1974	Polyethylene plant	68	High density	n.a.	n.a.	n.a.	

	Lab.)								
France	Creusot-Loire	1974	4 Ammonia plants	220	Ammonia	1979-80	100**	225**	At least 150 000 mt/y. to be shipped to France over 10-year period. Most to be sold in France and Europe. 40 000 mt/yr. of ammonia to Rhone-Poulenc. ** refers to shipments during 1975-80 only; approx. 300 000 tonnes of ammonia will be shipped each year during the 10-year life of the agreement.
Italy	ENI	1974 (preliminary agreement)	6 Chemical plants	(135) (110-165)*	Chemical products	n.a.	n.a.	n.a.	* Figure is based on the value of the 1 plant ordered so far; the total value of all 6 plants is given as $670-1 000 million.
United States	Occidental Petroleum	1974	2 Ammonia plants; fertilizer storage and handling facilities; ammonia pipeline	5 000	Ammonia, urea, potash	1978	5 000 (1978-1987)		10-year agreement; part of 20-year agreement. 2.1 million t/yr. of ammonia for first 10 years; 2.5 million t/yr. of ammonia for next 10 years; 1 million t/yr. of urea and same of potash for 20 years.
Italy	Tecnimont	1975	Polypropylene	[115] 100-130[a]	Possibly chemical intermediates	n.a.	n.a.	n.a.	
Italy	ENI	1975	Urea plants	170	Possibly urea, ammonia	n.a.	n.a.	n.a.	
Italy	Pessindustria	1975	Surface-active detergent plant	8.3	Monoethylene-glycol, organic chemicals, surface active agents	n.a	n.a.	n.a.	
Germany/ United Kingdom	Klöckner Group & Davy Powergas GmbH	1976	Phthalic anhydride plant	50	Chemical products including phthalic anhydride, fumaric acid and urea	1981[a]	47 over 10 yr. period (1981-90)		Mixed counterpurchase and buy-back.
France	Rhône-Poulenc S.A.	1976 (preliminary agreement)	Chemical plant	1 300[b]	Petroleum products and ammonia	1980	34.5 (1981-1990)		Sov. prods. to Rhône-Poulenc for in-house use. Only $400 m. in actual contracts so far. Latest is 1977 Speichim $97 m. contract. Total value is scheduled to rise to $1 000 million in terms of the French export value; 30 000 t/yr. of methanol as part payment for this deal over 10 years.
France	Technip S.A.	1976	2 Petrochemical facilities	501.1	Petroleum ($400.8m) Resultant prod. ($100.3m)	1980	501.1	n.a	

Table A-1 *(cont'd)*.

Western Country	Western Supplier	Year Signed	Type of Soviet Import	Value of Soviet Imports (Million US$)	Type of Soviet Export	First Year of Soviet Deliveries	Value of Soviet Exports (Million US$)	Remarks
Germany/ United Kingdom	Davy Powergas (UK) ICI Klöckner (Germany)	1977	2 Methanol plants	250	Methanol	1981-91	350[a] n.a.	Contract in dollars not sterling. ICI and Klöckner to purchase methanol approx. 20% of plants prod. or 300 000 mt/yr.
Germany	Klöckner	1975	PVC plant	40	PVC	1977	n.a. n.a.	10 000 t.p.a. for 10 years of vinyl chloride from 1977, 8 000 t.p.a. for 10 years of PVC from 1980.
United States	Lummus-Monsanto	1975	Acetic Acid plant	200	Acetic Acid	n.a.	n.a. n.a.	
Germany	Krupp-Koppers GmbH and Friedrich Uhde GmbH	1976	Dimethylterephthalate complex and addition to an existing polyester fibre plant	124.7	Raw cotton and base materials of aromatics	n.a.	n.a. n.a.	Possible $1.1 billion private credits. Hermes guarantee of credit facilities. Soviet goods mainly to German importers.
Germany	Hoechst	1976	De-emulsifiers for oil exploration, synthetic fibres and mat. textile dyes, and pigments, plant protection, pharm. preparations	130[a]	n.a.	n.a.	n.a. n.a.	Soviets now requesting 100% counter-trade for deals over $50 m.
Germany	Salzgitter	1976	Ethylene oxide plant	100	Resultant chemical product	1979		
Germany	Hoechst-Uhde-Wacker	1974	VCM plant capacity: 270 000 t/yr.	47	VCM	1976	66 (1976-1979)	40 000/50 000 tons of VCM over 4 years.
Germany	Klöckner-Hills	1974	PVC plant capacity: 250 000 t/yr.	71	VCM and PVC		33	10 000 t/yr. of VCM over 10 years. But deliveries not started yet. Also PVC buy-back.
Germany	Klöckner-Hills	1974	PVC plant capacity: 250 000 t/yr.	71	PVC and VCM	1978	54 (1978-1987)	8 000 t/yr. of PVC over 10 years (rest paid in VCM).
Germany	Krupp, Korf, Salzgitter, Siemens, Demag	1975	Production for pelleting of iron ore	1 000		1976	450 1 000	
				100	MEG	1979	100	27 000 t/yr. of MEG over 10 years.

60 000 t/yr.

Germany	Krupp-Koppers	1976- 1978	2 DMT plants: 60 000 and 120 000 t/yr. capacity	75 125	DMT	1981 1981	100 (1981-1990) 150 (1981-1990)	DMT taken as buy-back with other products.
Germany	Uhde-Hoechst	1977	Polyester staple fibre plant: capacity 35 000 t/yr.	70	Methanol	1981		Methanol as buy-back (with DMT and paraxylene).
Germany	Klöckner-Davy Power Gaz	1977	Phtalic anhydride plant plus maleic anhydride plant: capacity: 60 000 t/yr. and 3 000 t/yr.	50	Phtalic anhydride and maleic anhydride	1980	50 (1980-1989) (without urea)	5 000 t/yr. of phtalic anhydride and 3 000 t/yr. of maleic anhydride (plus urea buy-back).
United Kingdom/ United States	John Brown Contractors/ Union Carbide	1974	Polyethylene plant; capacity: 200 000 t/yr. high-density polyethylene	40	High-density polyethylene	1980	70 (1980-1989)	10 000 t/yr. of high-density polyethylene.
United Kingdom/ United States	John Brown Contractors/ Union Carbide	1977	Polyethylene plant; capacity: 200 000 t/yr. high-density polyethylene	90	Polyethylene	1983	162 (1983-1993)	Total of 240 000 tons of low-density polyethylene probably over 10 years.
Italy	Snia Viscosa	–	Caprolactam plant; capacity: 80 000 t/yr.	180	Caprolactam		224	28 000 t/yr. of caprolactam over 8 years.
Italy	ENI	1975	Chemical plants	1 000		1976		(for 1975-1980).
Italy	ENI	1975	2 plants	(170)				
Italy	Snamprogetti/Anic	1975	3 urea plants; capacity: 1 500 t/day each	200	Ammonia	1979	115 (1979-1988)	100 000 t/yr. of ammonia over 10 years.
Italy	Montedison	1975	Chemical plants	500		1977	175 250	
Japan	Marubeni	1975	Extension of plant to 75 000 t/yr. of acrylonitrile	10	Acrylonitrile	1978	30 (1978-1982)	10 000 t/yr. of acrylonitrile over 5 years.
Japan	Mitsui-Toyo	1976	4 ammonia plants; capacity: 1 360 t/day each	90	Ammonia and urea	1977	240 (1977-1987)	100 000 ton/yr. of ammonia and 100 000 ton/yr. of urea for 10 years.
France	Litwin	1974	Chemical plants	100		1977	50 60	
France	Creusot-Loire	1975	Chemical plants	220		1979	100 225	
France	Creusot-Loire	1975	Ammonia	200	Pipe			

Table A-1 *(cont'd)*.

Western Country	Western Supplier	Year Signed	Type of Soviet Import	Value of Soviet Imports (Million US$)	Type of Soviet Export	First Year of Soviet Deliveries	Value of Soviet Exports (Million US$)	Remarks
France	Technip S.A.	1976	2 aromatic complexes; capacity: 125 000 t/yr. of benzene; 165 000 t/yr. of orthoxylene; 165 000 t/yr. of paraxylene		Orthoxylene, paraxylene, benzene, diesel, oil, naphta	1980	950 (1980-1989) and (1981-1990)	20 000 t/yr. of orthoxylene, 20 000 t/yr. of paraxylene, 50 000 t/yr. of benzene; also diesel oil and naphta over 10 years.
France	Rhône-Poulenc	1976	Chemical plant	1 203				See also remarks for Rhône-Poulenc deal, in this Table.
United Kingdom/ Germany/ France	Woodal-Duckham, TBA-Bishop, Klöckner,INA	1979	Glass fibre plant	36	Methanol or ethylene glycol	–	–	For commission at Polotsk in 1982. Part of a 950 million credit line negotiated in 1975.
				Natural Resource Development				
Austria	Voest Oemv	1969	Large-diam. pipe	100	Natural gas	1969	900 1 000	Similar agreements signed in 1974 and 1975.
Finland	n.a.	1970	Pipe	n.a.	Natural gas	1974	n.a. n.a.	
Germany	Ruhrgas, Mannesmann Export AG	1970 1972 1974	Large-diam. pipe	1 500	Natural gas	1974	2 800 4 700	Three separate deals.
Italy	ENI	1971	Large-diam. pipe	190	Natural gas	1974	1 200 3 200	
France	Gaz de France	1972	Gas field equipment	250	Natural gas	1976	700 1 462	
France	CMP (incl. a subs. of International Systems and Control Corp.)	1974	Filter separators and gas compressor stations	26	Natural gas	n.a.	n.a. n.a.	
Italy	Finsider	1974	Large-diam. pipe	1 500	Scrap metal, coal, iron ore	1975	n.a. n.a.	* based on 65% coverage of Soviet import cost by Soviet exports. (Credit terms and costs are unknown and hence excluded.)
Japan	Southern Yakutsk	1974	Coal development	450	Coal	1979	80 860	Soviets have requested additional credits

France/ Austria/ Germany	Unspecified firms	1976	Large-diam. pipe and equipment	900[a]	Natural gas	1981	n.a.	n.a.	Part of triangular deal with Iran with participation of French and Austrian interest.
Finland	n.a.	1976	Construction of a mining complex (Kostamush)	n.a.	Iron pellet	n.a.	n.a.	n.a.	Series of counterdeliveries over unspecified time.
					Other				
United States	Airco. Inc.	1971	1 500 ton/yr. steel mill	n.a.	Chromium ore	n.a.	15	(for 1976)	This agreement is being extended to include welding equipment for super conduc. materials.
Belgium	Société Pachon	1976	Carpets	10.82	Textile raw materials and petroleum prod.	1977	12.85		
Finland	Rauma-Repola Oy	1977	3 tankers	n.a.	Ten 3 500 HP engines	1977-79	n.a.	n.a.	
Germany	Olympia-Werke	n.a. (1977)	Electric typewriter plant	40	Typewriters	1977	n.a.	n.a.	
United States	Philip Morris (via Eur. sub.)	1977	Tobacco production machinery	n.a.	Tobacco	1978	n.a.	n.a.	Reciprocal value agreement.
Germany	Glahe International	1976	Exhibition hall built with German labour and materials	31.1	n.a.	n.a.	n.a.	n.a.	Glahe will charge rentals to West. users for a period of 10 years after which hall will revert to USSR.
Germany	Rueterbau of Hannover-Langenhagen	1977	Air terminal at Moscow Airport	93.2	Energy and construction material	n.a.	n.a.	n.a.	
United States	Pepsi Cola	1974	Equipment, soft drink concentrates	n.a.	Vodka	n.a.	n.a.	n.a.	In 1976, Pepsi Cola announced it will supply equip. for add. cola plants. Equip. del. to start in mid-1977. Payment made in Vodka.
France	Péchiney-Ugine-Kuhlmann Group	1976	Alumina refinery	250	Aluminium	1979	n.a.	n.a.	USSR to sell back 50 000 tons of aluminium products from a mill in Siberia, also being constructed by Pechiney over 8 ½ years.
Finland	Kalmet	1976	Floating hotels; timber carriers; roll on/roll off vessels; ships	322	12 000 to capacity floating dock	n.a.	n.a.	n.a.	
United Kingdom	Dumbee-Combex Marx	1976	Toy moulds	n.a.	Toys	n.a.	n.a.	n.a.	10-year agreement. Soviet toys to be sold outside CMEA.

Table A-1 *(cont'd).*

Western Country	Western Supplier	Year Signed	Type of Soviet Import	Value of Soviet Imports (Million US$)	Type of Soviet Export	First Year of Soviet Deliveries	Value of Soviet Exports (Million US$)		Remarks
United States	FMC	1977	Tomato paste factory	n.a.	Tomato paste	n.a.	n.a.	n.a.	
Denmark	Danish Co-operative Union	1977	Clothing and canned hams	1.5	Oil prod.	n.a.	1.5	n.a.	
France	Péchiney-Ugine-Kuhlmann Group	1976	Aluminium complex (refinery)	(800) 600-1 000	Aluminium	n.a.	n.a.	n.a.	Agreement provides for the counterdelivery of 50 000 tonnes of aluminium products from another plan over 8 ½ years.
Japan	Yokohama, Aiwa, Tairiku Trading Companies	1976	Tires, steel rope, consumer goods	40	Timber, seafood	1976	60	n.a.	
Potential Agreements with the West									
Japan/ United States	US Japanese firms in competition	U.N.	Pulp-paper plants	1 000-2 000	Wood products	Early 1980s	n.a	n.a	
United Kingdom	Price & Pierce Ltd.	U.N.	Wood processing plant	120-220	Forestry products	n.a.	n.a.	n.a.	
United States/ Japan	Tokyo Gas, El Paso Co., Occidental Petroleum	U.N.	Development of natural gas deposits in Eastern Siberia	100	Liquified natural gas	1980s	n.a.	n.a.	
United States	Tenneco Inc., Texas Eastern Brown & Root Inc.	U.N.	Development of natural gas deposits in Western Siberia – North Star project	3 000[a]	Liquified natural gas	1980s	n.a.	n.a.	
France	Parfums Christian Dior	U.N.	Know-how for cosmetic production and packaging	n.a.	Unspecified items for use in cosmetics	n.a.	n.a.	n.a.	
France	Creusot-Loire	U.N.	300 000 ton/yr. steel pipe manufacturing plant	300	Unspecified related equipment	n.a.	n.a.	n.a.	
United States	Bendix Corp.	U.N.	Construction of spark plug	n.a.	Spark plugs	1980s	n.a.	n.a.	Production: 50-75 million plugs, of which 25 per cent to be sold in West by Bendix

Germany	Philips Petroleum, Montedison								Germany: aromatics likely specialised.
Japan	Nippon Electric Co.	U.N.	Electronic equip.	n.a.	n.a.	n.a.	n.a.	n.a.	Nippon willing ot accept 30% counterpurchase of Mash-Promintorg products.
Japan	Consortium	U.N.	4 pressurised water reactors and equipment	432	Enriched uranium	n.a.	n.a.	n.a.	
United Kingdom	Alfred Dunhill, Ltd.	U.N.	Cigarette factory	n.a.	Cigarettes	n.a.	n.a.	n.a.	20-30% of value of contract to be taken in product; Dunhill willing to take Soviet cigarettes for one-third of its national market.
United Kingdom	Cadbury-Schweppes Overseas, Ltd.	U.N.	Drinks technology and concentrate	n.a.	n.a.	n.a.	n.a.	n.a.	Payment in hard currency and rubles with Cadbury willing to use this for counterpurchase.
United States	Colgate-Palmolive Co.	U.N.	Powdered Laundry det. plant	n.a.	Powdered laundry detergent	n.a.	n.a.	n.a.	Partial compensation in kind; also counterpurchase of other products.
Japan	Taiyo Fishery Co.	U.N.	Refrig. plants, fishnet mfg. plants, canneries	34	Alaska pollack, whalemeat and other fis				
Germany	German Consortium (preliminary agreement)	U.N.	Steel complex	1 000-2 000[a]	Iron or pellets	1979	450	1 000	Initial contract and agreement signed in 1975, for cash only. Second stage may bring total value of Soviet imports to at least $2 billion and will include compensation in result product.

n.a. = Not available.
U.N. = Under negotiation.
a) Estimated.
b) Value for whole deal including several contracts.
Source: Eugene Zaleski and Helgard Wienert, *Technology Transfer between East and West* (Paris: Organisation for Economic Co-operation and Development, 1980), Table A-33, pp. 364-371, compiled from sources cited there.

Table A-2. **Examples of licences bought by the URSS from Western firms**

Description of the Technology	Western Licensor[a]	Announced
	Automotive	
Togliattigrad automotive plant – Positork automatic ignition device	DBA (Fr)	1/76
	Business Equipment	
Electric typewriters	Olympia Werke (FRG) (announced July 1974)	
	Chemicals and Petrochemicals	
Aromatics	Arco Chemical (US)	11/72
Chloroprene monomer on butadiene base	BP Chemicals International (UK)	3/73
Reinforced plastic foil	Ewald Dörken (FRG)	8/73
Alpha calcium-sulphate semihydrate refining	Gebr. Giulini (FRG)	9/74
High solid latex	International Synthetic Rubber (UK)	3/73
Acetic acid	Lummus Co and Monsanto	12/73
Automatic zinc-removing devices used in electrolysis	Montedison (I)	12/72
Isocyanate processing	Upjohn Co (US)	10/72
200 cm reactor for production of suspension PVC	Chemische Werke Huels AG (FRG)	4/75
"Pattex" contact glue	Henkel and Co (FRG)	5/75
Polymerization agent Liladox, a percarbonic acid derivative	Kemanord (Swe)	7/75
"Betanal", a herbicide for turnip and beet fields	Schering AG (FRG)	5/75
Porous material for acetylene bottles	L'Air liquide (Fr)	7/76
Synthesised standard gases	Seitetsu Kagaku Kogyo (Ja)	3/77
	Construction	
Roadbuilding and paving equipment	CMI Corp. (US)	10/76
	Consummer Goods	
Stainless steel razors	Wilkinson Sword (UK)	8/73
Padlocks and mortise locks	Wärtsilä (Fin)	8/76
Photoflash cubes	Bellmann (FRG)	11/76
	Electrical equipment	
Air preheaters for power stations	Kraftanlagen Heidelberg (FRG)	2/73
Axial bellows for power static cauldrons	Kühnle, Kopp & Kausch (FRG)	8/72
Cassette magnet head	Wolfgang Bogen (FRG)	5/74
High-voltage powerline insulation materials	General Cable (US)	2/77
	Electronics	
Automatic line for reed relays	Wm. Günther (FRG)	7/77
Thermistors plant	Murata Manufacturing (Ja)	8/77
	Food Products and Tobacco	
Marlboro cigarettes	Philip Morris (US)	2/77
	Houseold equipment	
Phonograph cabinets	Berlin Consult (FRG)	1/74
Electric stoves	Merloni SpA (I)	9/73
	Iron and Steel	
Conversion coating of cold rolled steel strips	Amchen products (UK)	9/72
Direct reduction process to be used in Kursk furnace	Midrex Corp. (US)	4/75
Steel structure manufacturing plant	Blohm & Voss (FRG)	1/77

Table A-2 *(cont'd)*

Description of the Technology	Western Licensor [a]	Announced
	Machine-Tools	
Wedge presses and related transport equipment	Eumuco (FRG)	12/73
Abrasive material	Norton (US)	1/73
Universal presses	Aïda Engineering (Ja)	7/74
	Materials-Handling Equipment	
Containers	Renault Industries Équipements et Techniques (Fr)	12/73
	Medical Equipment	
Disposable plastic medical goods	Portex (UK)	8/77
	Metalworking	
Aluminium wire	W.C. Heraeus (FRG)	8/77
	Mining and Metallurgy	
Aluminium casting; manufacturing equipment	Péchiney-Ugine-Kuhlmann (Fr)	11/76
	Packaging	
Nylon film production plant	Kohjin (Ja)	6/76
	Petroleum and Gas	
Ethylbenzene	Universal Oil Products	1/74
Gas dessication Orenburg natural gas complex	Davy Power Gas (FRG)	3/76
Oil drilling platform	Armco International (US)	7/76
Offshore exploitation of gas and oil, including blowout preventers, preventer control devices, Sea King and Marine Riser systems	Seitetsu Kagaku (Ja)	5/77
	Printing	
Two-web offset presses	Maschinenfabrik Augsburg-Nüremberg (FRG)	9/74
	Pulp and Paper	
Know-how and equipment for production of "Super Perga" paper	Greaker Industrier (No)	5/75
	Rubber	
Butadiene-type poly-chloroprene rubber	DuPont de Nemours (US)	8/74
	Shipping and Shipbuilding	
Pipe-sealing technology	Chuetsu-Waukesha (Ja)	6/77
	Textiles, Clothing and Leather	
Yield-increasing raw wool scouring	Sover S.A. (Be)	1/74
Clothing factory	MacIntosh Confectie (N)	1/77
Corset tulle	Gold-Zack Werke (FRG)	8/77

Table A-2 *(cont'd).*

Description of the Technology	Western Licensor[a]	Announced
Not Distributed by Industrial branch		
Porous acetylene bottles	L'Air Liquide (Fr)	
Axis-blower for nuclear power stations	A.G. Kühnle, Kopp & Kausch (FRG)	
Chemical treatment of steel strips	Anchem Products (UK)	
Furnaces for sulphur burning	Chemibau Zieren (FRG)	
Numerically controlled machine-tools	Fujitsu (Ja)	
Modular switches	Isostat (Fr)	
Motor vehicle brakes	Knorr-Bremsen (FRG)	
Machine-tool heads	Line (Fr)	
Resistors and equipment for their manufacture	Précis (Fr)	
Coating of metal sheets for motor vehicles	Pro Finish Metals (US)	
Prefabricated houses	Tchersmachiner (Swe)	
Electro-hydraulic cranes	Xegglound and Sioner (Swe)	

a) Country abbreviations: A : Austria; Aul: Australia; Be: Belgium; Fin: Finlande; Fr: France; FRG: Federal Republkic of Germany; I: Italy; Ja: Japan, N: Netherlands; No: Norway; Swe: Sweden; Swi: Switzerland; UK: United Kingdom; US: United States.

Source: Eugene Zaleski and Helgard Wienert, ***Technoly Transfer between East and West*** (Paris, Organisation for Economic Co-operation and Development, 1980), Table A-29, pages 343-346, compiled from sources cited there.

Table A-3. **Value of OECD Industrial goods exports to the USSR, by branch of origin, 1970-1982**[a]

Millions of US dollars

Branch of Origin	1970	1971	1972	1973	1974	1975	1976	1977	1978	1979	1980	1981	1982
1. Products of mining													
Crude minerals	2.0	1.9	2.5	3.0	2.8	7.5	9.4	12.2	17.3	15.4	15.1	59.1	90.0
Metalliferrous ores	7.1	5.3	5.6	7.7	7.4	18.4	19.4	40.7	79.0	113.1	72.1	137.7	109.5
Coal	..	0.0[b]	..	0.0	0.0	0.1	0.0	0.0	0.0	0.0	0.1	0.2	0.1
Petroleum and gas	..	0.0	..	0.0	0.0	0.1	0.0	0.1	0.2	0.0	0.0	0.0	0.0
Total	9.1	7.1	8.1	10.7	10.1	26.0	28.9	53.0	96.5	128.6	87.4	197.1	199.6
2. Coke and manufactured gas	0.0	0.0	0.0	0.0	0.1	0.0	0.0	0.0	0.0	0.0	0.3	0.0	0.0
3. Energy	0.0	0.0	0.0	0.0	0.0	0.0	0.0	0.1	0.1	0.3	0.0	..	0.0
4. Food processing													
Products of milling industry	0.0	0.0	0.1	0.0	0.0	0.0	1.1	1.8	1.5	0.5	108.1	332.7	153.6
Other food processing; beverage industry	4.1	6.2	9.4	10.7	50.0	69.3	47.2	36.1	36.4	94.2	109.4	132.3	80.1
Tobacco manufactures	1.6	1.1	0.7	0.6	1.0	1.0	1.4	1.8	2.9	1.6	4.1	45.0	1.9
Meat	42.7	50.1	2.3	10.2	138.9	105.1	108.7	259.7	30.8	278.7	449.5	531.7	307.6
Dairy products	15.1	14.8	29.6	118.6	35.5	36.3	46.0	93.6	61.1	253.9	407.7	272.0	313.6
Other processed food	28.3	25.6	58.8	44.1	108.1	129.4	187.1	181.7	203.4	295.6	764.6	1 163.4	888.8
Total	91.7	97.7	100.9	184.2	333.4	341.0	391.5	574.7	336.2	924.6	1 843.3	2 477.2	1 745.6
5. Textiles													
Knitted products	37.2	30.6	48.2	41.3	85.0	140.9	116.9	87.4	47.7	93.3	132.9	136.1	110.9
Other textiles	9.1	1.2	28.9	20.8	33.9	20.8	5.4	9.2	15.8	29.8	16.3	14.0	15.5
Manufactured textiles	137.7	163.4	173.0	146.2	233.0	246.1	248.7	322.2	246.3	338.2	574.9	508.4	454.6
Total	184.0	195.2	250.1	208.3	351.9	407.8	371.0	418.8	309.9	461.4	724.1	658.6	581.1
6. Clothing	99.0	78.3	75.8	72.0	101.5	152.7	133.9	138.4	127.6	112.9	257.3	412.4	319.9
7. Leather, shoes and furs	59.8	54.2	62.1	84.7	85.2	150.4	118.3	207.6	159.5	209.8	245.8	279.4	275.4
8. Paper, pulp, and processed wood													
Processed wood, except furniture	3.1	2.2	2.8	9.9	6.8	13.2	16.8	19.4	18.9	27.6	51.3	48.3	137.0
Furniture	5.7	7.1	10.9	9.5	15.6	15.1	15.5	17.4	20.6	34.1	49.7	56.8	48.9
Paper and products	184.4	190.3	203.6	221.4	356.2	614.1	471.0	521.5	499.4	589.8	1 024.8	977.9	926.5
Total	193.2	199.6	217.3	240.8	378.5	642.4	503.3	558.3	538.9	651.5	1 125.8	1 083.0	1 112.3
9. Products of printing industry	7.6	7.2	9.5	10.5	14.6	25.6	23.8	23.9	31.4	37.7	52.5	48.1	44.5

Table A-3 *(cont'd)*.

Branch of Origin	1970	1971	1972	1973	1974	1975	1976	1977	1978	1979	1980	1981	1982
10. Glass and china													
Glass and glass products	3.8	3.0	3.6	3.4	5.1	7.2	10.1	14.9	12.5	19.7	18.5	19.0	20.7
China	3.1	7.7	11.2	2.8	12.2	10.8	7.8	8.6	8.3	7.7	8.3	6.1	8.1
Total	6.9	10.7	14.8	6.1	17.3	18.0	18.0	23.5	20.8	27.4	26.9	25.1	28.8
11. Chemicals													
Basic industrial chemicals	104.2	119.8	124.1	141.3	317.9	413.5	430.5	710.8	785.3	1 101.3	1 333.8	1 104.0	984.8
Fertilizers and pesticides	13.2	10.9	25.7	15.8	41.1	73.5	48.8	64.5	73.6	121.0	198.8	125.0	152.3
Synthetic resins, plastic and man-made fibres	106.4	109.7	110.2	116.0	332.0	341.1	286.5	270.0	299.9	372.3	700.8	612.7	508.4
Paints and varnishes	11.6	12.8	11.9	10.7	24.9	23.7	24.9	26.8	38.6	53.3	100.9	125.9	92.1
Drugs, medicines	6.2	5.8	7.6	8.4	11.7	15.3	18.2	16.5	21.5	26.7	43.7	56.4	66.2
Soaps, perfumes	19.2	15.2	13.8	15.2	11.7	27.2	21.0	25.6	37.6	103.8	221.8	224.8	106.9
Gasoline, oils, other petroleum products	4.6	6.9	6.8	9.3	23.0	36.0	36.6	49.1	63.7	70.8	109.0	171.5	273.9
Other products of petroleum and coal	0.3	0.0	0.0	0.1	0.6	2.5	4.5	4.1	16.4	21.6	14.6	23.2	16.5
Rubber products	6.6	16.0	16.3	19.8	37.4	43.8	30.1	42.0	48.7	61.1	104.4	141.3	105.3
Other chemicals	19.7	83.4	26.7	32.2	61.7	95.7	103.2	99.1	135.3	177.2	233.3	213.9	197.1
Total	292.1	380.4	343.2	368.6	861.9	1 072.2	1 004.4	1 308.4	1 520.5	2 109.1	3 061.0	2 798.7	2 503.5
12. Other non-metallic mineral products	5.9	7.2	12.6	13.5	23.8	31.2	52.4	51.5	60.4	70.0	61.2	69.1	73.6
13. Metallurgy													
Iron and steel	297.7	368.6	488.5	975.0	1 981.5	2 554.4	2 790.8	2 277.0	2 858.6	3 741.9	3 628.2	3 523.0	4 138.0
Non-ferrous metals	32.7	17.7	20.7	31.2	39.8	42.3	76.5	88.9	184.9	214.6	265.2	239.2	208.5
Total	330.4	386.3	509.2	1 006.2	2 021.3	2 596.7	2 867.3	2 365.9	3 043.5	3 956.5	3 893.4	3 762.1	4 346.5
14. Machinery, equipment, and metal products													
Engines and turbines	0.9	1.6	6.2	13.2	22.1	78.1	59.6	236.7	214.2	42.5	21.7	21.1	107.8
Agricultural machinery	25.5	18.1	4.5	31.6	11.7	133.1	177.2	19.1	63.5	64.9	117.8	170.5	116.2
Metal and woodworking machinery	236.9	200.0	394.3	579.8	705.9	1 048.5	1 026.6	1 157.6	1 196.3	1 160.2	1 216.4	846.2	1 056.6
Textile machinery	26.5	15.9	40.0	59.6	88.2	206.0	164.8	93.5	96.2	123.9	110.3	122.4	134.8
Paper and pulp machinery	23.4	18.1	17.2	35.7	48.3	32.4	73.2	76.7	76.8	117.5	39.7	30.6	57.3
Printing and bookbinding machinery	15.2	15.9	11.4	20.6	17.6	18.3	18.5	20.0	34.2	28.7	26.0	47.6	37.3
Food processing machines	2.5	3.5	8.2	8.7	18.7	25.0	21.4	15.5	19.5	28.3	34.5	31.5	22.2
Construction and mining machinery	25.1	17.7	11.4	9.8	20.4	127.5	137.4	100.1	121.5	136.9	132.3	326.8	554.0
Glass working machinery	5.1	3.7	6.1	9.6	11.1	8.3	7.1	6.5	7.4	21.1	20.9	16.2	3.0
Other machinery for special													

machinery	26.7	33.4	40.0	42.6	41.8	86.6	86.5	81.8	113.7	139.9	116.5	140.1	[illegible]
Electrical industrial machinery and apparatus	19.3	23.1	24.6	25.6	45.8	75.5	101.9	170.0	279.8	222.2	144.3	116.9	175.4
Radio, TV, and communication equipment	25.1	18.1	24.9	31.9	43.8	62.9	65.0	90.8	117.1	134.0	126.2	120.0	126.3
Electrical appliances and housewares	0.3	1.0	0.9	0.8	0.7	1.1	1.5	1.9	2.4	3.0	3.4	6.6	5.5
Ships	192.8	115.3	64.2	100.0	180.3	485.7	549.1	575.2	680.4	711.3	511.7	621.4	761.2
Railroad equipment	5.7	11.5	11.9	18.8	23.7	17.9	26.0	25.1	39.8	20.1	30.7	32.5	36.9
Motor vehicles	30.3	37.6	44.8	29.6	55.2	430.8	403.0	171.0	177.6	219.7	313.9	369.6	613.9
Motorcycles and cycles	0.0	0.1	0.2	0.2	0.2	0.3	0.7	0.2	0.1	0.2	0.2	0.1	0.1
Aircraft	0.5	0.3	0.3	1.0	0.9	6.4	0.9	1.7	2.2	0.1	2.2	0.2	0.9
Scientific, measuring, and control equipment	57.2	57.1	70.2	78.1	110.4	174.1	176.9	220.9	241.8	321.7	311.4	273.5	277.4
Photographic and optical goods	8.8	12.1	8.8	8.2	11.5	18.9	16.1	20.3	20.4	33.4	31.6	24.9	30.7
Watches and clocks	0.3	0.2	0.4	0.9	0.2	0.4	0.6	0.8	1.8	1.3	1.3	1.8	1.0
Structural metal products	10.2	15.8	32.5	30.8	61.8	105.4	210.8	307.3	258.1	231.6	157.3	185.4	272.8
Cutlery and hand tools	20.5	16.2	16.3	17.2	26.2	50.0	50.9	67.6	62.3	69.4	60.3	66.2	80.7
Other electrical apparatus	38.4	34.3	40.6	49.2	118.3	202.8	144.6	242.8	259.9	308.3	351.7	330.5	283.4
Total	888.9	756.3	1 011.5	1 420.9	1 930.3	3 854.5	4 001.6	4 224.1	4 659.7	4 656.8	4 341.9	4 166.1	5 238.7
15. Other industrial commodities	263.6	223.0	370.5	504.0	734.0	1 251.3	1 666.0	2 062.2	2 074.7	1 892.7	2 172.1	1 879.8	2 105.9
16. Total industrial exports	2 432.3	2 403.2	2 985.6	4 130.4	6 864.1	10 569.8	11 180.2	12 010.5	12 979.7	15 239.3	17 893.0	17 856.6	18 575.4

a) The classification corresponds to the USSR branch of origin classification. Components may not add to totals because of rounding.
b) Less than 0.1 million US dollars.
Source: East-West Technology Transfer Data Base, OECD, Paris.

Table A-4. **Deflated value of OECD industrial goods exports to the USSR, by branch of origin, 1970-1981**[a]

Millions of 1970 US dollars

Branch of Origin	1970	1971	1972	1973	1974	1975	1976	1977	1978	1979	1980	1981
1. Products of mining												
Crude minerals	2.0	1.8	2.6	2.7	1.8	2.7	3.7	4.5	7.1	9.8	8.9	22.4
Metalliferrous ores	7.1	7.2	8.2	6.6	4.7	9.9	16.3	36.7	80.7	65.6	29.5	88.4
Coal	..	0.0[b]	..	0.0	0.0	0.0	0.0	0.0	0.0	0.0	0.0	0.1
Petroleum and gas	0.0	0.0	..	0.0	0.0	0.0	0.0	0.0	0.0	0.0	0.0	0.0
Total	9.1	9.0	10.9	9.3	6.5	12.6	20.0	41.2	87.8	75.4	38.4	111.0
2. Coke and manufactured gas	0.0	0.0	0.0	0.0	0.0	0.0	0.0	0.0	0.0	0.0	0.1	0.0
3. Energy	0.0	0.0	0.0	0.0	0.0	0.0	0.0	0.0	0.0	0.1	0.0	..
4. Food processing												
Products of milling industry	0.0	0.0	0.1	0.0	0.0	0.0	0.8	1.5	1.4	0.4	59.1	164.9
Other food processing; beverage industry	4.1	12.6	14.3	17.7	82.0	159.3	45.8	127.9	108.4	290.8	330.7	336.4
Tobacco manufactures	1.6	0.9	0.6	0.5	0.6	0.5	0.6	0.8	1.3	0.7	1.8	18.6
Meat	42.7	49.0	2.4	6.3	54.4	45.9	52.8	125.1	19.8	130.1	142.2	159.0
Dairy products	15.1	14.9	22.2	199.6	25.8	21.8	32.9	78.8	39.3	169.4	202.5	101.8
Other processed food	28.3	31.7	73.4	25.2	58.8	66.7	140.8	137.7	119.5	149.3	314.1	419.4
Total	91.7	109.2	113.0	249.3	221.6	294.3	273.7	471.7	289.7	740.6	1 050.4	1 200.1
5. Textiles												
Knitted products	37.2	32.6	45.5	27.4	42.7	87.6	66.1	46.7	25.6	48.2	60.6	75.9
Other textiles	9.1	1.3	31.8	21.5	19.8	12.9	3.9	5.7	7.1	11.1	6.0	5.1
Manufactured textiles	137.7	166.6	163.4	97.0	117.0	153.1	140.6	172.1	132.4	174.8	261.9	283.6
Total	184.0	200.5	240.7	145.9	179.5	253.7	210.5	224.5	165.1	234.2	328.5	364.6
6. Clothing	99.0	71.4	60.7	45.4	58.6	78.1	66.9	61.5	48.0	37.4	83.8	160.3
7. Leather, shoes and furs	59.8	50.0	56.6	68.8	70.2	119.3	92.0	137.4	106.0	123.0	104.6	131.0
8. Paper, pulp, and processed wood												
Processed wood, except furniture	3.1	5.7	23.7	45.9	22.6	25.0	35.5	43.5	73.1	66.5	78.4	147.8
Furniture	5.7	6.6	9.0	6.3	9.5	8.4	8.5	8.5	8.4	12.3	16.7	22.3
Paper and paper products	184.4	182.8	180.4	179.5	192.5	230.6	252.0	274.8	267.6	280.2	427.2	429.9
Total	193.2	195.1	213.0	231.7	224.6	264.0	296.0	326.8	349.2	359.0	522.3	600.0
9. Products of printing industry	7.6	6.2	7.0	5.9	7.8	11.4	10.7	9.8	10.9	11.4	15.1	16.0
10. Glass and china												
Glass and glass products	3.8	2.9	3.6	2.8	4.1	3.9	6.1	7.2	6.2	7.2	8.9	7.1
China	3.1	7.4	11.2	2.3	9.9	5.8	4.7	4.2	4.0	2.8	3.6	2.3
Total	6.9	10.3	14.8	5.1	14.0	9.7	10.8	11.3	10.2	10.0	12.5	9.4
11. Chemicals												
Base industrial chemicals	104.2	118.3	148.4	132.7	129.5	189.1	341.0	328.4	399.6	354.7	426.3	416.4
Fertilizers and pesticides	13.2	10.5	23.5	14.2	24.2	29.2	25.6	28.8	41.5	50.3	75.7	55.0
Synthetic resins, plastic and man-made												

Soaps, perfumes	19.2	15.8	16.3	11.0	12.3	20.9	29.4	23.5	30.2	59.8	112.3	148.4
Gasoline, oils, other petroleum products	4.6	2.8	1.6	3.9	1.8	5.3	4.6	5.2	8.2	6.8	12.3	17.9
Other products of petroleum and coal	0.3	0.0	0.0	0.0	0.5	0.4	1.4	1.2	2.9	2.9	2.6	2.9
Rubber products	6.6	12.4	12.0	11.1	15.3	19.7	14.7	20.7	23.5	23.0	35.6	66.1
Other chemicals	19.7	22.8	23.7	23.0	38.7	49.4	60.1	50.6	66.8	68.5	84.9	95.3
Total	292.1	331.1	372.3	342.2	437.6	525.8	711.1	666.5	801.8	856.7	1 189.8	1 263.6
12. Other non-metallic mineral products	5.9	7.0	12.7	11.3	19.4	16.8	31.5	24.8	30.0	25.5	29.8	25.6
13. Metallurgy												
Iron and steel	297.7	393.4	482.0	742.4	1 040.7	1 052.2	1 416.1	1 021.3	1 784.8	1 508.3	1 431.9	1 586.3
Non-ferrous metals	32.7	18.3	18.6	23.5	21.2	23.7	48.4	46.3	100.8	77.5	104.4	156.2
Total	330.4	411.8	500.5	765.9	1 061.9	1 075.9	1 464.5	1 067.6	1 885.6	1 585.7	1 536.3	1 742.5
14. Machinery, equipment, and metal products												
Engines and turbines	0.9	1.4	4.7	8.2	12.2	37.2	27.4	96.2	72.3	12.4	6.1	7.0
Agricultural machiney	25.5	15.6	3.4	20.0	6.5	65.8	84.0	7.9	22.4	19.7	33.8	56.7
Metal and woodworking machinery	236.9	177.0	306.4	356.8	379.0	501.4	480.7	480.2	417.6	353.1	347.3	286.3
Textile machinery	26.5	13.9	30.9	36.1	49.5	101.5	78.9	39.5	34.7	39.6	32.8	43.2
Paper and pulp machinery	23.4	16.3	13.5	21.8	25.5	14.7	32.4	29.5	24.9	33.4	10.6	9.7
Printing and bookbinding machinery	15.2	14.3	8.9	12.6	9.3	8.3	8.2	7.7	11.1	8.2	7.0	15.1
Food processing machines	2.5	3.0	6.1	5.0	9.7	11.3	9.4	6.1	6.1	7.8	9.0	9.7
Construction and mining machinery	25.1	15.5	8.6	5.9	10.6	56.1	58.3	37.2	37.9	37.2	35.4	105.4
Glass working machinery	5.1	3.3	4.9	6.1	6.2	4.0	3.3	2.7	2.6	6.8	5.7	5.3
Other machinery for special industries	91.4	77.1	104.6	138.2	147.6	234.4	245.9	238.6	223.8	179.5	150.5	100.7
Office and computing machinery	26.7	31.5	34.7	30.6	26.3	49.8	52.4	43.9	53.9	59.2	53.4	81.5
Electrical industrial machinery and apparatus	19.3	21.3	20.3	17.3	26.0	37.9	50.5	74.6	105.5	74.8	45.6	45.7
Radio, TV, and communication equipment	25.1	17.3	22.1	23.0	29.1	37.6	39.3	47.4	57.7	61.4	56.1	65.6
Electrical appliances and housewares	0.3	0.9	0.8	0.5	0.4	0.6	0.8	0.9	1.0	1.1	1.2	2.9
Ships	192.8	96.4	46.3	55.3	96.0	232.4	261.6	231.1	212.1	222.1	151.1	220.7
Railroad equipment	5.7	9.7	8.4	9.9	12.8	8.7	13.8	9.6	15.3	8.4	11.7	12.2
Motor vehicles	30.3	33.2	34.7	18.2	31.3	211.8	193.0	71.0	62.2	68.4	92.3	128.0
Motorcycles and cycles	0.0	0.1	0.1	0.1	0.1	0.1	0.3	0.1	0.0	0.1	0.1	0.0
Aircraft	0.5	0.2	0.2	0.6	0.5	3.1	0.4	0.7	0.7	0.0	0.6	0.1
Scientific, measuring, and control equipment	57.2	50.9	55.5	49.3	64.9	91.0	89.9	98.5	90.4	105.8	96.4	98.5
Photographic and optical goods	8.8	11.0	7.1	5.2	6.5	9.6	8.0	8.9	7.5	10.9	10.2	9.6
Watches and clocks	0.3	0.2	0.3	0.6	0.1	0.2	0.3	0.4	0.7	0.5	0.5	0.8
Structural metal products	10.2	13.2	18.4	14.3	25.0	31.9	70.0	91.0	72.2	52.1	30.3	80.4
Cutlery and hand tools	20.5	13.5	8.3	7.7	10.3	14.1	15.2	18.2	16.9	14.5	11.2	28.8
Other electrical apparatus	38.4	32.1	33.5	32.4	67.1	101.2	70.6	103.9	93.3	97.4	107.4	118.8
Total	888.9	668.7	782.8	875.7	1 052.6	1 864.6	1 894.8	1 746.0	1 643.1	1 472.2	1 306.3	1 533.0
15. Other industrial commodities	263.6	255.8	266.9	293.7	377.2	547.3	717.6	794.8	693.8	573.9	641.8	683.6
16. Total industrial exports	2 432.3	2 326.1	2 652.1	3 050.3	3 731.7	5 073.5	5 800.1	5 584.0	6 121.2	6 107.2	6 859.6	7 840.5

a) The classification corresponds to the USSR branch of origin classification. Components may not add to totals because of rounding.
b) Less than 0.1 million US dollars.
Source: East-West Technology Transfer Data Base, OECD, Paris.

Table A-5. **Value of OECD capital goods exports to the USSR, by type of product, 1970-1982**[a]

Millions of US dollars

Type of Product	1970	1971	1972	1973	1974	1975	1976	1977	1978	1979	1980	1981	1982
1. Stationary power plants and water engineering													
Stationary power plants	6.5	9.5	20.3	26.0	44.7	117.1	130.0	312.9	328.8	131.3	67.5	64.9	177.8
Water turbines and engines	0.5	0.5	0.9	1.3	2.9	3.2	4.3	17.8	1.5	3.5	5.8	6.5	9.0
Total	7.0	10.0	21.2	27.2	47.6	120.3	134.3	330.7	330.3	134.8	73.4	71.4	186.9
2. Electric power distribution	37.9	38.1	38.7	40.8	83.7	130.9	140.8	245.5	296.9	290.2	274.3	224.7	219.7
3. Liquid fuel, gas, and water distribution													
Tubes and pipes	160.7	95.5	108.1	273.7	351.7	792.9	1 092.5	903.0	1 039.4	1 351.3	1 129.8	1 400.2	1 940.1
Pumps and centrifuges	21.2	24.6	54.7	75.9	99.8	171.3	248.4	351.4	500.2	371.3	306.4	171.5	230.6
Total	181.8	120.1	162.8	349.6	451.5	964.3	1 340.9	1 254.4	1 539.6	1 722.7	1 436.2	1 571.7	2 170.7
4. Transport equipment													
Rail transport equipment	0.4	0.5	0.4	6.0	7.2	27.9	11.1	12.2	14.8	6.4	8.2	19.0	5.8
Road transport equipment	13.4	25.6	32.5	11.2	27.1	318.3	309.7	99.5	81.3	97.9	140.0	141.8	389.0
Aircraft	0.0[b]	0.0	0.0	0.0	0.4	5.2	0.1	0.0	0.0	0.0	0.1	0.0	0.0
Sea and river transport equipment	181.2	101.5	50.5	86.3	162.4	447.8	510.1	534.8	631.1	650.2	437.1	545.0	664.7
Total	195.0	127.7	83.5	103.6	197.1	799.1	830.9	646.5	727.2	754.5	585.3	705.7	1 059.5
5. Agricultural equipment													
Field machinery	0.6	0.5	1.3	0.9	1.1	6.9	5.9	1.5	4.2	12.0	7.4	14.4	23.2
Dairy farm equipment	0.4	0.5	0.7	3.0	1.0	1.3	6.5	2.3	4.2	5.8	4.3	5.7	16.5
Tractors	23.8	14.8	2.0	26.9	7.8	121.9	165.5	14.8	53.8	44.4	96.3	142.5	74.4
Presses for making wine, juices, etc.	0.0	0.0	0.0	0.0	0.5	0.1	0.1	0.3	0.2	0.3	0.8	0.3	1.0
Other agricultural machinery	0.8	2.2	0.5	0.8	1.4	2.8	2.7	0.4	1.0	2.3	9.1	7.7	1.0
Total	25.5	18.1	4.6	31.6	11.7	133.1	180.7	19.3	63.5	64.9	117.8	170.5	116.2
6. Construction and mining machinery	30.9	25.7	29.1	39.7	78.1	211.3	207.9	168.5	224.0	252.5	181.9	370.9	595.7
7. Engineering, welding, and metallurgical equipment (excluding furnaces)													
Machine tools	173.8	117.3	244.5	362.9	448.3	550.5	576.9	650.7	637.5	635.8	749.6	470.8	524.3
Metallurgical machinery	17.8	25.7	9.7	11.8	54.7	80.0	74.0	84.1	100.7	70.7	23.1	22.5	17.3
Welding appliances and equipment	0.2	0.2	0.5	0.7	2.6	3.2	7.2	9.2	8.5	6.1	8.6	5.9	7.5
Total	191.8	143.2	254.7	375.4	505.6	633.7	658.0	744.0	746.7	712.5	781.3	499.3	549.0
8. Industrial and laboratory furnaces and gas generators	53.4	30.6	37.7	51.0	78.7	125.3	153.9	185.5	144.0	139.5	138.6	100.2	122.2

11. Electronics and telecommunications													
Telecommunications	17.0	7.0	10.9	16.5	20.8	23.9	28.1	30.3	57.0	69.9	51.3	54.9	63.0
Computers	18.6	22.1	23.9	26.9	21.9	43.6	52.3	39.4	68.7	79.3	61.9	88.6	56.0
Optical instruments	2.2	2.7	2.7	4.3	4.3	8.8	6.2	11.0	10.0	14.1	11.8	12.9	18.3
Other electronics	0.2	0.1	0.2	0.1	0.0	0.2	0.5	0.0	2.2	0.6	0.1	3.0	0.0
Total	38.0	32.0	37.8	47.8	47.1	76.5	87.1	80.7	137.9	164.0	125.1	159.4	137.3
12. Machinery for special industries													
Textile and leather machinery	29.2	17.6	44.2	62.3	95.0	210.9	173.9	107.5	99.7	128.9	120.1	135.8	153.6
Pulp and paper mill machinery	23.4	18.1	17.2	35.7	48.3	32.4	73.2	76.7	76.8	117.5	39.7	30.6	57.3
Food processing machinery	12.8	16.6	25.1	33.4	52.8	74.6	73.8	63.7	66.7	99.8	132.3	92.2	97.5
Glass working machinery	5.1	3.7	6.1	9.6	11.1	8.3	7.1	6.5	7.4	21.1	20.9	16.2	3.0
Printing and bookbinding equipment	15.2	16.0	11.4	20.6	17.6	18.3	18.5	20.0	34.2	28.7	26.0	47.6	37.3
Total	85.7	72.0	104.1	161.6	224.8	344.5	346.5	274.4	284.8	396.0	339.0	322.4	348.7
13. Mechanical handling equipment and storage tanks													
Mechanical handling equipment	45.6	58.6	42.6	118.5	122.3	259.7	237.1	267.7	235.2	248.6	316.7	350.6	467.4
Storage tanks	0.9	3.0	5.7	5.6	7.3	13.0	23.2	43.9	58.4	43.8	10.8	13.1	24.7
Total	46.5	61.6	48.3	124.1	129.5	272.7	260.3	311.5	293.6	292.4	327.5	363.8	492.0
14. Office machines	1.0	0.9	1.6	1.8	1.8	2.8	3.1	2.5	4.2	4.9	4.3	4.5	2.9
15. Medical apparatus, instruments, and furniture	5.5	8.4	8.8	12.4	17.8	27.1	32.6	33.0	37.8	42.5	54.4	58.2	51.3
16. Heating and cooling of buildings and vehicles	2.1	2.3	2.7	2.7	4.4	5.6	11.8	22.1	22.2	18.3	19.6	47.8	31.2
17. Measuring, controlling, and scientific instruments	52.6	50.7	63.1	68.5	98.0	154.0	156.5	187.6	202.3	262.5	262.8	223.7	227.8
18. Finished structural parts and structures	8.8	11.4	19.2	21.5	50.2	83.9	149.2	212.0	177.7	151.6	134.7	165.4	242.5
19. Other capital equipment													
Other non-electrical machinery	97.7	99.0	141.7	227.0	277.9	493.7	499.4	545.4	613.8	559.8	481.1	289.5	386.7
Other equipment	39.4	45.7	78.6	64.6	84.8	214.5	403.2	608.6	536.4	411.2	286.0	177.5	234.8
Total	137.1	144.7	220.3	291.6	362.8	708.3	902.6	1 154.0	1 150.2	971.0	767.2	467.0	621.5
Grand total	1 126.4	922.8	1 222.8	1 886.6	2 495.3	4 918.3	5 663.4	6 006.7	6 519.2	6 546.9	5 712.9	5 607.6	7 261.6

a) Components may not add to totals because of rounding.
b) Less than 0.1 million US dollars.
Source: East-West Technology Transfer Data Base, OECD, Paris.

Table A-6 **Value of OECD capital goods exports to the USSR, by end use, 1970-1982**[a]

Millions of U.S. dollars

End Use	1970	1971	1972	1973	1974	1975	1976	1977	1978	1979	1980	1981	1982
Industry	617.0	545.0	873.7	1 201.6	1 614.9	2 635.7	2 946.1	3 664.1	3 824.0	3 513.6	3 041.5	2 559.9	3 269.8
Oil and gas distribution	175.6	113.6	149.6	337.8	432.8	919.9	1 279.3	1 179.7	1 398.5	1 648.1	1 389.9	1 522.4	2 099.2
Agriculture	25.5	18.1	4.6	31.6	11.7	133.1	180.7	19.3	63.5	64.9	117.8	170.5	116.2
Construction	0.0(b)	0.0	0.0	0.0	0.0	0.0	0.0	0.0	0.0	0.2	0.3	0.0	0.2
Domestic trade	0.0	0.0	0.0	0.0	0.0	0.1	0.1	0.0	0.0	0.1	0.0	0.0	0.0
Transport	194.9	127.7	83.3	102.2	193.1	789.9	812.7	658.9	722.4	743.2	568.2	694.9	1 030.7
Communications	17.0	7.0	10.9	16.5	20.8	23.9	28.1	30.3	57.0	69.9	51.3	54.9	63.0
Administration and management	19.6	23.0	25.5	28.7	23.7	46.4	55.4	41.9	72.9	84.2	66.2	93.1	58.9
Public health	5.5	8.4	8.8	12.4	17.8	27.1	32.6	33.0	37.8	42.5	54.4	58.2	51.3
Non-attributable	71.6	80.7	67.2	157.0	182.3	344.7	329.8	381.6	346.4	383.6	427.7	458.5	578.2
Total	1 126.8	923.6	1 223.6	1 887.8	2 497.3	4 920.8	5 664.8	6 008.7	6 522.5	6 550.4	5 717.2	5 612.3	7 267.4

a) Components may not add to totals because of rounding.
b) Less than 0.1 million US dollars.
Source: East-West Technology Transfer Data Base, OECD, Paris.

Table A-7. **Deflated value of OECD capital goods exports to the USSR, by end use, 1970-1981**[a]

Millions of 1970 US dollars

End Use	1970	1971	1972	1973	1974	1975	1976	1977	1978	1979	1980	1981
Industry	61.70	485.8	676.8	742.8	877.7	1 250.7	1 353.1	1 484.1	1 324.6	1 068.7	871.7	897.5
Oil and gas distribution	175.6	117.8	138.0	247.5	230.7	389.5	644.2	522.9	780.8	641.6	525.1	673.6
Agriculture	25.5	15.7	3.5	20.0	6.5	65.8	85.5	7.9	22.4	19.7	33.8	56.7
Construction	0.0[b]	0.0	0.0	0.0	0.0	0.0	0.0	0.0	0.0	0.1	0.1	0.0
Domestic trade	0.0	0.0	0.0	0.0	0.0	0.0	0.1	0.0	0.0	0.0	0.0	0.0
Transport	194.9	107.2	61.0	56.1	103.2	382.5	389.1	265.4	227.8	233.6	168.0	246.9
Communications	17.0	6.9	10.0	12.4	14.3	15.5	18.7	18.6	30.6	34.8	26.1	34.4
Administration and management	19.6	21.8	22.6	21.1	15.6	28.3	35.2	24.3	36.0	38.0	33.2	58.6
Public health	5.5	7.7	7.2	8.0	10.5	14.2	16.5	14.4	14.2	14.0	17.4	22.1
Non-attributable	71.6	70.3	51.1	93.1	98.7	161.8	152.9	154.2	116.7	113.6	119.0	152.4
Total	1 126.8	833.1	970.2	1 201.0	1 357.3	2 308.2	2 695.2	2 491.9	2 553.3	2 164.1	1 794.4	2 142.2

a) Components may not add to totals because of rounding.
b) Less than 0.1 million US dollars.
Source: East-West Technology Transfer Data Base, OECD, Paris.

Table A-8. **Value of OECD industrial capital goods exports to the USSR, by industrial end-use branch, 1970-1982[a]**

Millions of US dollars

End-Use Branch	1970	1971	1972	1973	1974	1975	1976	1977	1978	1979	1980	1981	1982
1. Mining and fuel extraction machinery	26.3	19.7	12.8	13.3	23.0	132.0	140.3	103.2	126.4	141.8	136.2	332.7	562.1
2. Raw materials processing industry	5.8	8.0	17.8	29.9	57.7	83.8	70.4	68.5	102.5	115.6	49.6	44.1	41.8
3. Energy-power plants													
Steam and vapour	0.8	2.2	12.8	6.8	9.4	9.9	52.0	92.0	40.5	48.8	15.7	9.1	7.6
Gas turbines	0.0[b]	0.0	0.0	8.4	14.1	73.9	40.2	176.1	190.9	25.5	8.8	12.3	95.1
Nuclear reactors	0.0	0.0	0.0	0.0	0.0	0.0	0.3	0.0	0.7	0.3	0.4	..	0.0
Electric power machinery	5.6	7.3	7.5	10.7	21.2	33.2	37.4	44.4	95.3	56.2	42.1	42.7	74.4
Other	0.4	0.2	0.3	0.4	1.5	0.9	0.5	2.5	1.5	3.5	5.8	6.5	9.0
Total	6.9	9.7	20.6	26.3	46.3	117.9	130.4	315.0	329.0	134.2	72.9	70.6	186.2
4. Food processing and tobacco industry													
Milling	0.3	0.8	0.5	0.4	3.9	1.5	0.6	0.9	0.9	9.8	14.8	11.8	2.2
Other	14.4	17.8	29.9	36.0	54.9	93.9	78.6	75.6	66.2	90.8	117.6	80.6	95.8
Total	14.7	18.6	30.3	36.4	58.8	95.4	79.2	76.5	67.1	100.6	132.5	92.4	98.0
5. Textile industry													
Spinning and extruding	4.2	4.1	2.2	14.3	22.2	99.5	63.1	28.8	5.9	21.0	8.1	8.0	33.7
Weaving and knitting	7.9	6.6	25.6	28.5	22.4	45.2	52.0	24.6	30.0	36.1	32.3	37.0	31.8
Felt manufacturing	0.1	0.0	1.1	1.9	4.1	3.2	0.2	0.5	0.0	2.5	5.0	9.4	0.8
Other	8.8	4.7	8.0	10.7	28.7	42.0	35.7	32.4	56.2	62.1	59.3	49.6	60.3
Total	21.0	15.4	37.0	55.4	77.4	189.9	150.9	86.3	92.1	121.7	104.8	104.0	126.7
6. Clothing industry	2.7	1.7	4.8	3.6	7.7	5.6	10.0	15.0	3.4	5.0	9.8	13.5	18.8
7. Leather, shoe, and fur industry	5.5	0.4	2.4	3.2	10.0	15.5	13.0	6.3	4.1	2.2	5.5	18.4	8.2
8. Paper, pulp, and wood-processing industry													
Paper and pulp	23.7	18.9	18.1	37.0	50.3	34.9	74.6	78.7	80.1	121.0	44.1	35.4	63.2
Machine tools for wood-processing	10.6	10.7	63.1	111.6	53.3	62.4	26.6	30.6	23.5	49.1	28.3	35.8	33.6
Total	34.4	29.6	81.2	148.6	103.6	97.4	101.3	109.3	103.6	170.1	72.4	71.1	96.8
9. Printing industry													
Bookbinding	0.3	1.4	0.8	7.5	5.0	3.9	2.7	3.2	5.3	6.8	5.2	9.8	8.0
Typesetting	2.9	2.2	3.0	4.3	3.7	4.7	3.4	6.8	5.2	4.9	2.5	5.5	9.4
Other	12.1	12.4	7.6	8.9	8.9	9.7	12.4	10.0	23.7	17.0	18.3	32.3	19.9
Total	15.2	16.0	11.4	20.6	17.6	18.3	18.5	20.0	34.2	28.7	26.0	47.6	37.3
10. Glass and china industries	6.1	7.8	12.4	15.9	17.5	16.0	19.2	13.2	23.1	42.3	37.6	25.7	20.6
11. Engineering and metal working													

Furnace burners	0.5	0.9	1.8	2.6	3.6	2.0	4.9	9.1	2.6	2.9	4.9	5.1	3.2
Other	45.2	24.2	26.5	38.2	60.7	103.3	79.5	105.3	73.6	88.3	97.7	79.3	103.4
Total	63.5	50.8	38.0	52.6	119.0	185.3	158.3	198.5	176.9	161.9	125.7	106.9	123.8
13. Coke and gas industry	7.8	5.5	9.3	10.2	14.4	20.0	69.6	71.1	67.8	48.3	36.1	15.9	15.6
14. Chemicals and construction materials	45.8	61.3	88.4	74.5	98.3	236.6	471.5	698.8	733.9	528.7	312.4	216.9	313.1
15. Non-attributable													
Metal structural parts	8.8	11.4	19.2	21.3	50.1	80.7	147.8	210.4	175.4	147.1	134.5	165.4	242.4
Electricity distribution equipment	84.4	79.0	95.1	98.2	178.1	280.1	259.8	439.0	531.1	566.7	489.9	413.2	403.7
Machinery and mechanical appliances	91.4	85.8	131.8	217.4	265.2	458.6	481.0	520.9	570.6	516.9	457.6	263.6	365.7
Other	3.2	7.0	16.7	11.3	21.9	52.4	48.0	61.5	45.2	45.9	88.4	87.1	84.9
Total	187.7	183.2	262.8	348.2	515.4	871.7	936.6	1 231.9	1 322.3	1 276.6	1 170.4	929.3	1 096.6
Grand total	617.0	545.0	873.7	1 201.6	1 614.9	2 635.7	2 946.1	3 664.1	3 824.0	3 513.6	3 041.5	2 559.9	3 269.8

a) Components may not add to totals because of rounding.
b) Less than 0.1 million US dollars.
Source: East-West Technology Transfer Data Base, OECD, Paris.

Table A-9. **Deflated value of OECD industrial capital goods exports to the USSR, by industrial end-use branch, 1970-1981[a]**

Millions of 1970 US dollars

End-Use Branch	1970	1971	1972	1973	1974	1975	1976	1977	1978	1979	1980	1981
1. Mining and fuel extraction machinery	26.3	17.3	9.7	8.1	12.1	58.3	59.7	38.5	39.6	38.6	36.6	107.5
2. Raw materials processing industry	5.8	7.2	14.1	18.8	32.4	39.8	32.9	28.4	36.4	37.4	13.6	14.4
3. Energy-power plants												
Steam and vapour	0.8	1.9	9.7	4.1	5.1	4.9	24.6	38.0	13.9	15.0	4.5	3.2
Gas turbines	0.0[b]	0.0	0.0	5.3	7.9	35.2	18.8	72.6	64.7	7.5	2.5	4.1
Nuclear reactors	0.0	0.0	0.0	0.0	0.0	0.0	0.2	0.0	0.2	0.1	0.1	..
Electric power machinery	5.6	6.7	6.2	7.2	12.0	16.6	18.5	19.4	35.9	18.9	13.3	16.7
Other	0.4	0.1	0.2	0.2	0.9	0.4	0.2	1.0	0.5	1.0	1.7	2.2
Total	6.9	8.8	16.1	16.9	25.9	57.2	62.3	131.0	115.3	42.5	22.2	26.2
4. Food-processing and tobacco industry												
Milling	0.3	0.7	0.3	0.2	2.0	0.7	0.3	0.3	0.3	2.7	3.9	3.6
Other	14.4	16.0	23.1	21.9	29.7	44.0	36.5	31.0	22.7	26.6	33.7	27.4
Total	14.7	16.7	23.4	22.1	31.7	44.6	36.7	31.3	23.0	29.3	37.6	31.1
5. Textile industry												
Spinning and extruding	4.2	3.6	1.7	8.7	12.4	49.0	30.2	12.1	2.1	6.7	2.4	2.8
Weaving and knitting	7.9	5.7	19.8	17.2	12.6	22.3	24.9	10.4	10.8	11.5	9.6	13.1
Felt manufacturing	0.1	0.0	0.9	1.1	2.3	1.6	0.1	0.2	0.0	0.8	1.5	3.3
Other	8.8	4.2	6.2	6.5	16.1	20.7	17.1	13.7	20.3	19.8	17.6	17.5
Total	21.0	13.5	28.6	33.6	43.4	93.6	72.2	36.4	33.2	38.9	31.1	36.7
6. Clothing industry	2.7	1.6	3.8	2.3	4.5	2.9	5.2	7.0	1.3	1.7	3.2	5.2
7. Leather, shoe, and fur industry	5.5	0.4	1.9	2.0	5.6	7.6	6.2	2.6	1.5	0.7	1.6	6.5
8. Paper, pulp, and wood-processing industry												
Paper and pulp	23.7	16.9	13.9	22.3	26.3	15.4	32.8	30.1	25.8	34.1	11.4	11.8
Machine tools for wood-processing	10.6	9.7	49.5	68.7	29.2	30.1	12.7	12.8	8.2	14.6	8.3	12.5
Total	34.4	26.6	63.4	91.0	55.4	45.5	45.5	42.9	34.0	48.8	19.7	24.3
9. Printing industry												
Bookbinding	0.3	1.2	0.6	4.5	2.7	1.7	1.2	1.2	1.7	1.9	1.4	3.1
Typesetting	2.9	1.9	2.4	2.6	1.9	2.1	1.5	2.6	1.7	1.4	0.7	1.7
Other	12.1	11.2	5.9	5.4	4.7	4.4	5.5	3.9	7.7	4.8	4.9	10.3
Total	15.2	14.3	8.9	12.6	9.3	8.3	8.2	7.7	11.1	8.2	7.0	15.1
10. Glass and china industries	6.1	7.1	9.8	9.9	9.7	7.7	9.1	5.5	8.1	13.1	10.6	8.6

Furnace burners	0.5	0.8	1.4	1.6	2.0	0.0	2.2	3.6	0.9	0.9	1.3	1.7
Other	45.2	21.3	20.5	23.5	34.0	50.2	37.3	43.4	25.4	27.0	29.2	27.7
Total	63.5	45.4	29.5	32.6	67.0	92.4	76.4	83.9	62.9	50.5	37.6	37.4
13. Coke and gas industry	7.8	4.9	7.3	6.4	8.0	9.1	31.1	28.1	22.4	14.1	9.9	5.2
14. Chemicals and construction materials	45.8	54.7	67.0	45.5	53.9	106.5	208.8	272.2	242.6	152.5	85.5	73.8
15. Non-attributable												
Metal structural parts	8.8	9.5	9.8	9.5	19.8	22.6	43.9	56.5	47.7	30.6	25.0	72.2
Electricity distribution equipment	84.4	72.4	77.4	64.5	103.2	144.0	131.0	193.5	198.4	187.8	153.9	153.5
Machinery and mechanical appliances	91.4	77.1	104.6	138.2	147.6	234.4	245.9	238.6	223.8	179.5	150.5	100.7
Other	3.2	6.1	12.9	7.1	12.2	25.8	23.1	26.2	17.0	15.1	27.1	31.6
Total	187.7	165.0	204.7	219.3	282.8	426.8	444.0	514.8	486.8	413.0	356.6	358.0
Grand total	617.0	485.8	676.8	742.8	877.7	1 250.7	1 353.1	1 484.1	1 324.6	1 068.7	871.7	897.5

a) Components may not add to totals because of rounding.
b) Less than 0.1 million US dollars.
Source: East-West Technology Transfer Data Base, OECD, Paris.

Table A-10. **Value of OECD exports of technology-based intermediate goods to the USSR, by type of product, 1970-1982**[a]

Millions of US dollars

Type of Product	1970	1971	1972	1973	1974	1975	1976	1977	1978	1979	1980	1981	1982
1. Parts and accessories of machinery and equipment	76.8	82.7	118.9	146.7	228.4	510.0	472.1	546.7	626.6	681.6	738.0	680.9	827.4
2. Parts and accessories of transport equipment	33.9	37.0	38.5	45.1	64.9	147.4	154.5	133.1	186.2	211.8	301.4	359.6	386.7
3. Paper manufactures	130.2	138.2	154.4	163.0	273.5	483.8	354.4	408.6	301.7	371.1	652.0	549.7	584.2
4. Textile manufactures	114.9	135.0	167.2	147.7	236.5	305.0	282.3	320.0	230.7	342.7	509.9	497.3	419.7
5. Synthetic rubber	3.6	6.5	3.2	5.4	23.2	38.5	30.4	25.5	24.7	37.9	59.2	81.0	52.8
6. Synthetic fibres	28.4	26.3	21.6	15.5	44.8	35.9	27.8	31.6	20.2	26.6	90.6	44.0	41.9
7. Special manufacture of leather, rubber, wood,glass, and minerals	0.7	2.7	5.6	11.9	13.9	27.6	15.5	32.8	30.2	35.0	29.5	23.0	16.5
8. Miscellaneous mineral manufactures	3.9	3.7	4.6	4.0	8.5	10.0	10.0	14.8	13.9	16.9	16.1	17.9	16.7
9. Iron and steel	137.1	273.0	380.4	701.0	1 629.7	1 760.1	1 697.9	1 373.9	1 818.5	2 386.3	2 498.2	2 122.7	2 197.7
10. Non-ferrous metals	32.0	17.7	19.9	30.7	39.5	41.3	72.1	84.7	177.7	213.3	257.4	234.6	202.3
11. Organic and inorganic chemical elements	100.7	117.7	113.6	134.3	290.0	384.0	408.8	666.9	687.5	1 004.6	1 182.7	988.7	847.7
12. Final chemical manufactures	26.6	27.8	28.4	26.8	66.8	63.1	59.5	72.1	88.6	113.9	179.2	199.1	138.1
13. Chemicals and plastic materials	120.8	119.9	138.9	142.8	376.6	448.9	379.0	401.6	483.4	628.1	1 097.6	909.0	838.2
14. Other intermediate goods	2.4	2.6	2.9	3.2	3.8	5.7	8.2	9.3	9.6	12.4	14.3	11.9	11.5
Total	811.8	990.8	1 198.2	1 578.2	3 300.2	4 261.4	3 972.4	4 121.6	4 699.5	6 082.3	7 626.0	6 719.4	6 581.3

a) Components may not add to totals because of rounding.
Source: East-West Technology Transfer Data Base, OECD, Paris.

Table A-11. **Examples of Western co-operation agreements with the USSR**

Western Partner	Soviet Partner	Purpose of Agreement	Announced
Agriculture			
Gi è Gi SAS (Italy)	Tractorexport	Production and marketing cattle-feeding complexes	1/74
Elanco (US)	Ministry of Agriculture	Joint tests of antibiotics for livestock breeding: exchange of results	10/75
E.I. DuPont de Nemours (US)	SCST*	Research, production and application of agricultural chemicals	8/77
Agricultural Equipment			
Robert Bosch (Germany)	SCST	Outfitting vehicles incl. tractors and agricultural machines, hydraulics and pneumatics, TV technology	12/73
Vereinigde Machinefabrieken (Netherlands)	SCST	Production of turbine blades, rotary screen printing machines, milk and products, offal processing in slaughterhouses	12/73
International Harvester (US)	Soyuselchoztechnika	Testing corn-tilling machinery	7/74
International Harvester (US)	SCST	R & D: manufacture of tractors and other farm and construction machinery	10/75
Aircraft			
Boeing (US)	SCST	R & D: helicopter and passenger aircraft construction, air traffic control	6/74
Rockwell International (US)	Aviaexport	Rockwell equipment assembled into Soviet planes for sale in the US	9/73
Lucas Aerospace Ltd.	SCST	Hydraulic and pneumatic aircraft control systems, onboard power-generating systems and switch gear, engine monitoring systems	5/75
Automotive			
Daimler-Benz (Germany)	SCST	Car production; R & D on safety, urban transport, pollution	3/73
Fiat SpA (Italy)		Expansion of Togliattigrad car works (negotiating)	
Marposs Finike Italiana		Automobile construction and fine mechanics, ball bearings; sales in third markets	2/75
Volvo (Sweden)	SCST	Production and R & D: combustion engines, electric motors	6/74
Daimler-Benz (Germany)	SCST	Automotive manufacturing	10/76
Fiat (Italy)	SCST	Renewed agreement to expand auto output and manufacture farm and building industry vehicles	11/76
Business Equipment			
Rank Xérox (UK)	SCST	R & D: copying and duplicating machines	6/73
Chemicals and Petrochemicals			
Agfa-Gevaert (Germany/Belgium)		R & D: photochemistry	4/73
BASF (Germany)	SCST	R & D: plastic materials, dyestuffs, fuels, mineral oil additives	8/72
Bayer (Germany)	SCST	Inorganic pigments, synthetic rubber and raw materials, lacquers, environmental protection	6/73
ENI (Italy)	Ministry of Chemical Industry	Construction of chemical plants against Soviet payments in kind	3/73

Table A-11 *(cont'd)*.

Western Partner	Soviet Partner	Purpose of Agreement	Announced
Klimsch & Co. (Germany)	SCST	R & D and modernisation of photographic reproduction equipment and technology	12/73
Liquichimica (Italy)	Ministries of Oil Refining and Petrochemicals and Chemicals	R & D: paraffins, olefins and derivatives; industrial biosynthesis; lubricating oil additives	7/74
Monsanto (US)	SCST	R & D: use of computers in chemical industry; development of rubber compound products	11/73
Montedison (Italy)	Foreign Trade Ministry	Construction of chemical plants against Soviet payments in products	8/73
Montedison (Italy)	Soyuschimexport	Barter: Italian base chemicals for Soviet chemical intermediates	6/74
Occidental Petroleum (US)		Construction of fertilizer plants	4/73
Pressindustrie (Italy)	SCST	Chemical and food industries, pharmacology and cosmetics	5/74
Reichhold Chemie (Austria)	SCST	Production of lacquers, resins and related raw materials	3/73**
		Production of synthetic polymeric materials, wood chemicals extraction	11/74
Schenectady Chemicals (US)	Ministry of Electrotechnical Industry	Enamel and varnish wire insulating techniques (protocol)	2/75
Snia Viscosa (Italy)	SCST	Artificial and synthetic fibers, environmental production	2/74**
Synres Nederland, Sigma (Netherlands); De Soto (US)	R & D Institute of the Paint Industry	Know-how on composition and use of paints in extreme climates	5/74
Rohm & Haas (US)	SCST	R & D in plastics, herbicides, pesticides and petroleum additives	5/75
Montedison SpA (Italy)	Licensintorg	Joint R & D of polycarbonates to serve as possible alloy, glass and ceramic substitutes	5/75
Phillips Petroleum (US)	Ministry of Oil Refining and Petrochemical Industry	Exchange of know-how of olefin industry porcesses	9/75
Hempel's Marine Paints (Denmark)	SCST	R & D: production of varnishes and paints; anti-corrosive coatings for ocean-going vessels	1/76
Krauss-Maffel (Austria)	SCST	Design of plastics processing equipment	6/76
Worthington Pumps Italiana	SCST	Joint production of pumps for chemical and petrochemical industries	6/76
	Ministry of Chemical and Petroleum Machine Building	Joint R & D of protein and chemical pumping equipment; exchange of experts and information	6/76
Imperial Chemical Industrie (UK)	SCST	Exchange of information on crop protection, chemicals, health products, chlorinated rubber, fibers, fertilizers, paints	6/76
L'Oréal (France)	SCST	Supplies of chemical products	7/76
Vianova Kunsthaerz (Austria)	SCST	Joint R & D; exchange of experience and experts in production and application of synthetic resins, varnishes, paints, enamels	8/76
Maschinenfabrik A. Gentil (Germany)	All-Union Research Institute of Hydromachinery Construction and Technology	Joint development of pumps for chemical industry	10/76
Mitsui Group (Japan)	SCST	Scientific and technical co-operation in chemical and petrochemical industries and electrical engineering	11/76

Table A-11 *(cont'd)*.

Western Partner	Soviet Partner	Purpose of Agreement	Announced
Chemie Linz (Austria)	SCST	Co-operation and joint R & D in plastics, pharmaceuticals, glues; exchange of experts and test results	12/76
Essochem Impex (Belgium)	SCST	Petrochemicals products	4/77
Ballestra (Italy)	SCST	Production of technical and medical white oils, sulphonates, detergents	7/77
Computers			
Burroughs (US)	SCST	Computer technology, inc. education, design, programming and application	8/74
Control Data (US)	SCST	R & D: computer technology	2/73
	SCST	R & D: computers for transportation, medicine and education	11/73
Sperry Rand (US)	SCST	Computer application in air traffic control and other areas, also the manufacture of farming machines	6/74
Construction			
Bechtel (US)	SCST	Control and organisation of planning and building large-scale industrial units	7/73
Manfred Swarovski GmbH** (Austria)	SCST	Co-operation in traffic safety	7/75
H.H. Robertson (US)	SCST	Exchange of information on rubber materials and dyes for structural coatings	8/75
Durisol (Switzerland)	SCST	Exchange of experiences in using lightweight building materials	8/75
Metecno (Italy)	SCST	Uses of prefab. panels in industrial and civil construction	3/76
Stetter (Germany)	SCST; Gosstroy	R & D: production of concrete mixing and setting equipment	9/76
Otis Elevator (US)	Ministry of Construction, Road and Municipal Machine-building		11/76
Consumer Goods			
Loewy-Raymond-Willion Snaith (US); Cie de l'esthétique industrielle (France)	All-Union Research Institute of Technical Aesthetics	Industrial and consumer design for cars, boats, household appliances, motorcycles	1/74
Revlon (US)	SCST	Joint R & D; exchange of staff for cosmetics, perfumes, fragrances, pharmaceuticals	2/76
Electrical Equipment			
General Electric (US)	SCST	Power generation, transmission and utilisation	1/73
Matsushita Electric Industrial Co. (Japan)	SCST	Electrical and electronic equipment	10/74
Siemens; Kraftwerk-Union (Germany)	Ministrye of Electrical Engineering	Computer testing, turbogenerators, water jackets, anticorrosion measures	7/73
General Electric's UK subsidiary	SCST	Development and engineering of turbogenerators, heavy machinery, and AC & DC transmission systems	5/75
Tokyo Boeki (Japan)	SCST	Scientific-technical co-operation and exchange of information on electrical technology	7/75
Maekawa Shoji (Japan)	SCST	Refrigeration; compressors	9/75
Gould (US)	SCST: Ministry of Electrotechnical Industry	Condensers, electro-motors, electronic tools, storage batteries, power metallurgy	12/75

Table A-11 *(cont'd).*

Western Partner	Soviet Partner	Purpose of Agreement	Announced
Alsthom (France)	Energomashexport	Hydro-accumulating power stations	6/76
Daikin Kogyo (Japan)	State Committee for Construction – Gosstroy	Energy-saving air-conditioning systems; exchange of technology	10/76
Siemens (Germany)	SCST	Electrical engineering	10/76
AEG-Telefunken (Germany)	SCST	Agreement renewed for joint production of household appliances	2/77
Hitachi (Japan)	Energomashexport	Construction and equipment of power stations in third countries	8/77
Jungner (Sweden)	Ministry of Electrotechnical Industry	Development of locomotive batteries	9/77
Electronics			
Arthur Andersen (US)		Application of control systems in industry, production calculations (protocol)	5/73
Hewlett-Packard (US)	SCST	Medical electronics, mini computers, measuring instruments	6/73
Nippon Electric (Japan)	SCST	R & D: communications and electronic technology; household electrical appliances	8/74
Olivetti SpA (Italy)**	SCST	Electronic and automation of production and management	7/75
Wolfgang Bogen (Germany)	SCST	Exchange information and experts in recording techniques	1/76
Marconi (UK)	SCST	Joint R & D on radio and TV equipment	2/76
AEG-Telefunken (Germany)	SCST	Electronics; radio engineering; precision instruments	10/76
CBS (US)	State Committee for Radio Broadcasting and Television	Exchanges programs and expertise; develop technical co-operation	10/76
Cameca (France)	Burevestnik	Jointly develop and produce microelectronic screen miscroscopes	4/77
Strömberg (Finland)	SCST	Computer control of cement production	7/77
Nippon Electric Co. (Japan)	SCST	Communications; electronic technology	7/77
W.C. Heraeus (Germany)	Ministry of Electronics Industry	Joint production of diascripters	8/77
Jungner (Sweden)	Ministry of the Electrotechnical Industry	Development of locomotive batteries	9/77
Environment			
Silvani Anticendi (Italy)	SCST	Firefighting technology	10/75
Food			
Coca-cola (US)	SCST	R & D in agricultural and food industry, including new foodstuffs with higher nutritional value; production of soft drinks; land reclamation methods	7/74
Pepsico (US)	Soyusplodimport	Pepsi-Cola marketing in USSR against buyback of Soviet liquors	12/72
Hennessy (France)	SCST	Food, wine, cosmetics	2/76
Rieber & Son (Norway)	Ministry of Fish Industry	Fish processing	3/77
Seagram (US)	SCST	Food processing	4/77
Gervais-Danone (France)	SCST	Joint R & D; exchange of information in food production, packaging, storage, marketing	7/77
Household Equipment			
Kymi Kymmne (Finland)	SCST; Ministry of the Construction Materials Industry	Plumbing and bathroom fixtures	2/77

Table A-11 *(cont'd)*.

Western Partner	Soviet Partner	Purpose of Agreement	Announced
Iron and Steel			
Fried. Krupp (Germany)	SCST	R & D: continuous casting methods	2/73
Lurgi-Gesellschaften (Germany)	SCST	Construction of metallurgical and chemical plants	2/74
Otto Wolf (Germany)	SCST	Production of rolling mills for third markets	2/74
Perusyhtyma Oy, Lemminkäinen Oy, Finnbothia (Finland)	Prommashimport	Mining iron ore against counterdeliveries of iron ore pellets	12/73
Schoeller-Bleckmann Stahlwerke (Austria)	SCST	R & D: rapid machining of high alloyed steel	7/73
Vöest (Austria)	SCST	R & D: converter steel technology	7/72
Vöest-Alpine (Austria)	Soyuzpromexport	Austrian imports of bituminous coal and iron ore	6/74
Vöest-Alpine (Austria)	Promsyrioimport	Export of sheet metal	6/74
Machines-Tools			
Association of French Machine-Tool Builders	Ministry of Machine Building and Instrument Making Industry	Automatic production lines for car spare parts; design of manufacturing equipment for wood packaging, saws	7/72
Elliot Co. (US)	Energomashexport	Production of compressors	12/73
Georg Fischer (Switzerland)	SCST	Engineering, casting techniques	7/73
Werkzeugmaschinen Fabrik Gildemeister (Germany)	SCST	R & D: machines-tools, metalworking machines	4/73
Sandvik (Sweden)	SCST	Joint R & D in metal-cutting tools, super-hard materials in metal processing	12/75
Gildemeister (Germany)	SCST	Machine-tools	10/76
Liebherr (Germany)	Ministry of Machine-Tools and Tool Building Industry	Grinding machines	4/77
Machinery			
Mitsubishi (Japan)	SCST	R & D: new machinery and equipment	4/74
Silvani Anticendi SpA (Italy)	SCST	Scientific-technological co-operation in fire-fighting	7/75
Ishikawajila-Harima Heavy Industries (Japan)	Ministry of Heavy, Power and Transport Machine-building	Design of rolling mills, furnace equipment, multipurpose cranes	7/76
Fried. Krupp (Germany)	SCST	Heavy machine building	10/76
Materials Handling			
Interpool Ltd. (UK) jointly with Associated Container Transportation (Australia), Nippon Yusen Kabushiki Kaisha and Yamashita Shinnihon Steamship (Japan)	Sovfracht	Leasing containers	2/73
Strick Co. (US)	Sovinflot	Leasing refrigerated containers	8/74
Medical Equipment			
Siemens (Germany)	Ministry of Health	R & D: Medical equipment, computerised diagnosis	7/74
LKB Produkter (Sweden)	SCST	Clinical biochemical instruments	6/76
Metalworking			
Occidental Petroleum (US)	Various FTOs	Barter: Metal-processing against Soviet raw nickel	12/72

Table A-11 *(cont'd).*

Western Partner	Soviet Partner	Purpose of Agreement	Announced
Degussa Deutsche Gold-und Silberscheideandstalt (Germany)	SCST	Scientific-technical co-operation in metalworking, participation in various Soviet industrial projects, including transfer of know-how and technology	7/75
Vereinigte Edelstahlwerke (Austria)	Paton Institute, Kiev	Technical application of electroslag remelting process	10/75
Mining			
Joy Manufacturing (US)		Machine building for coal and mining industries (protocol)	7/72
Kaiser Resources (Canada)	Ministry of Coal Industry	Hydraulic and open-cut strip mining	8/74
Ruhrkohle (Germany)	Ministry of Coal Industry; SCST	Design of mining equipment, safety systems; processing and preparation of residual products	5/73
Hermann Hemscheidt, Maschinenfabrik (Germany)	SCST; Ministry of Coal Industry	Joint R & D of machinery for working thin and medium coal seams	11/75
Ruhr Kohle (Germany)	SCST	Coal equipment and technology	10/76
Multisector			
AEG-Telefunken (Germany)	SCST	Data processing, energy, tool, communication, transportation	8/72
AKZO (Netherlands)	SCST	R & D: yarns, fibers, chemicals, pharmaceuticals, plastics and rubber products	8/72
Armco Steel Corp. (US)	SCST	Ferrous metalurgy, offshore drilling equipment	1/74
Bendix Corp. (US)	SCST	Automotive, aerospace and electronics products, scientific instruments, automation and machine tool products	11/74
Food Machinery Corp. (US)	SCST	Agricultural, food industry, packaging, oil industry and materials-handling equipment	12/73
General Dynamics (US)	SCST	Aviation industry, ship-building, communications	10/73
Gulf Oil (US)	Ministries of Geology, Coal, Oil Extracting, Chemical, Oil Refining and Petrochemicals	Exploration, production and transport of petroleum; processing petroleum and derivatives; production of chemicals and synthetic fuels	3/75
Industrial Nucleonics (US)	SCST	Automation and control in pulp and paper industry; steel, rubber and plastics manufacture	5/74
Kaiser Steel (US)	SCST	Aluminium production, special steel, large-diameter pipes, off-shore drilling platforms, coal hydro-extracting technologies	2/74
Linde AG (Germany)	SCST	R & D: cryogenic techniques, sewage treatment, natural gas refining, chemical equipment	5/74
Lockheed Aircraft (US)	SCST	Aviation industry, navigation systems and apparatus, special machine-tool building, cross-country vehicles, minicomputers, mineral exploration, medical electronic systems	2/74
Norton Simon (US)	SCST	Technology for cosmetics, soft drinks, production of baby food, food and packaging industry	10/74
Occidental Petroleum (US)	SCST	Extracting and processing oil and gas; agricultural fertilizers and chemicals; metalworking, metalcoating, projecting and building of hotels, utiisation of solid wastes	7/72

Table A-11 *(cont'd).*

Western Partner	Soviet Partner	Purpose of Agreement	Announced
Philips Morris (US)	SCST	Paper-packing materials, chemical technology, tobacco-leaf cultivation, cigarette manufacture	4/74
Singer Co. (US)	SCST	Calculators, training equipment, navigation equipment, consumer electrical products	10/73
Stanford Research Institute (US)	SCST	Scientific, technological and economic activities	10/73
Universal Oil Products (US)	SCST, Ministry of Oil Refining and Petrochemicals	Petroleum refining, organic chemicals and plastics technology (protocol)	12/74
Allis-Chalmers (US)	SCST; Ministries of Heavy, Power and Transport Engineering; Ferrous Metallurgy; and Non-ferrous Metallurgy	Exchange of know-how in engineering and metallurgy	4/75
Colgate-Palmolive (US)	Ministry of Chemical Industry	Medicine, sports goods and detergents	5/75
G.L. Rextroth GmbH, member of Mannesmann Group (Germany)		Co-operation in manufacture of control systems, especially for hydraulic excavators	5/75
Loewy Raymond International (US)	SCST	Joint design of packaging, industrial interiors, e.g. shopping centres and hotels	7/75
Nokia Oy (Finland)	SCST	Scientific-technical co-operation in electrical engineering, electronics rubber and paper production	5/75
Rolls-Royce Ltd (UK)	SCST	R & D in industrial engines and aerospace technology	5/75
Shell (UK)	SCST	Agrochemicals and oil drilling	5/75
Finmeccanica (Italy)	SCST	R & D: production of electrical and power-generating equipment, electronics rubber and paper production	10/75
Cie générale d'électricité (France)	SCST	Energy; petrochemical equipment; machine-tools	10/75
Union Carbide (US)	SCST	Chemicals; metallurgy, electric welding; environmental protection	10/75
Corning Glass (US)	SCST	Glass; glass ceramics; electronics; biochemistry	12/75
Ente Partecipazioni e Finanziamento Industria Manifatturiera (Italy)	SCST	Exchange of information and specialists; joint R & D in aluminium, shipbuilding, food processing, fishing industries	6/76
C. Itoh (Japan)	SCST	Electrical and electronic engineering; chemistry	7/76
Airco (US)	SCST	Manufacture of welding equipment; shipbuilding; medical equipment; refrigeration	11/76
Babcock & Wilcox (US)	SCST	Joint research	5/77
Plessey (UK)	SCST	Renewed agreement in aviation, telecommunications, data processing, semiconductors, hydraulic and pneumatic systems	6/77
Mitsubishi (Japan)	SCST	Extension of agreement in machine building, chemicals, petrochemicals, ferrous metallurgy, shipbuilding, engine manufacture	7/77
Non-ferrous Metals			
Klöckner-Humboldt-Deutz (Germany)	Licensintorg	Preparation and extraction of non-ferrous metals	3/74
Minemet (France)	Mekanobrabotka Institute	R & D: flotation units	2/73
Trefimétaux (France)	Gripros-Wetmetobrabotka Institute	R & D: high-resistant copper alloys	2/73

Table A-11 *(cont'd)*.

Western Partner	Soviet Partner	Purpose of Agreement	Announced
Nuclear			
Gulf Oil Corp. (US)	SCST	Nucler energy, atomic power	10/74
Kraftwerk Union (Germany)	SCST	Construction of nuclear power station	11/74
Packaging			
American Can (US)	Ministry of Engineering for Light and Food Industries	Container and packaging technology	1/73
	SCST	Container and packaging technology	7/74
Petroleum and gas			
Brown & Roots (US)	SCST	Engineering and construction of gas; oil transportation methods	6/73
Cooper Industries (US)	Licensintorg	Natural gas exploitation	9/72
Dresser Industries (US)	SCST	Geophysics research in oil and gas extraction	6/74
ENI (Italy)	Ministry of Gas Industry	Opening and exploiting gas deposits	4/72
Japanese consortium		Developing and exploiting oil and natural gas deposits on continental shelf of Sakhalin peninsula	2/75
Occidental Petroleum (US)	SCST	Oil and gas exploration	7/72
Petroleum Services, Subsidiary of Dresser Industries (US)	SCST	Private agreement involving only the subsidiary; oil and gas extraction	10/73
Worthington Pump (US)	Ministry of Chemical and Oil-Machine Building	Consultation and R & D on oil pipeline technology (protocol)	12/74
Phillips Petroleum (US)	Ministry of Oil Extraction Industry	Joint prospecting and exploitation of oil deposits; recovering oil as secondary crude from other industrial processes	9/75
Standard Oil of Indiana (US)	SCST	Oil prospecting and extraction; oil refining	9/75
British Petroleum (US)	SCST	Exchange of information on lubricants, bitumen, other petroleum by products; pollution control; synthetic proteins	6/76
Liquichimica (Italy)	Ministry of Oil Extraction Industry	Joint R & D in oil-exploiting technology	12/76
Essochem Impex (Belgium)	Ministry of Oil refining and Petrochemical Industry	Lube oil additives	4/77
Cameron Iron Works (US)	SCST	Exchange of specialists and information in petrochemical engineering	5/77
Pharmaceuticals			
American Home Products (US)	SCST	R & D: pharmaceuticals and medical instruments	9/74
Beecham Group (UK)	SCST	Exchange of documentation; R & D on antibiotics	2/73
Schering AG (Germany)	SCST	R & D	9/73
Pfizer Inc. (US)	SCST	Exchange of scientific-technical information, joint R & D of pharmaceuticals, particularly veterinary and botanic genetics	6/75
Revlon Inc. (EU)	Soyuzchimexport Raznoexport	Joint development and marketing of "Epas" perfume	5/75
Rosenlew Oil (Finland)	SCST	Scientific-technical co-operation in microbiological products	7/75
Abbott Laboratories (US)	SCST	Joint R & D in infant nutrition, drugs	11/75
	Ministry of Health; Ministry of the Meat and Dairy Industry	Execution of research	

Table A-11 *(cont'd).*

Western Partner	Soviet Partner	Purpose of Agreement	Announced
Bristol Myers (US)		Exchange of information in antibiotics, non-narcotic analgesics, chimotherapy in cancer research	9/76
Precision Instruments			
Lip (France)		Production of quartz-based wrist watch components	11/72
Rank Taylor Hobson (UK)	State Committee for Standards	Development of surface and optical metrology equipment for auto, machine-tool, aircraft and other industries	11/74
Varian Associates (US)	SCST	High energy particle accelerators for scientific, industrial and medical application, analytic and measuring instruments, vacuum apparatus and components	6/74
LKB Produkter (Sweden)	SCST	R & D, design and production of scientific instruments	1/76
Disa Elektronik (Denmark)	SCST	Production of scientific instruments	8/77
Pulp and Paper			
International Paper (US)	Ministry of Pulp and Paper	R & D in paper industry and on plant construction	6/74
Parsons & Whittemore (France)	Prommashimport	Construction of pulp plant against counter-deliveries of pulp	8/73
Printing			
BASF (Germany)	SCST	Expansion of agreement to include photopolymeric printing plates	8/76
Railway			
Plasser & Theurer (Austria)	SCST	Rail laying, maintenance, and track tamping	7/73
SKF (Germany)	Soviet railway authorities	Joint development of a railway wheel bearing	7/77
Rubber and Products			
Continental Gummiwerke (Germany)	SCST	Know-how for rubber mixing factory	9/74
Dunlop-Pirelli (UK/Italy)	SCST	R & D: car tire	11/72
Services			
McKinsey & Co. (US)	SCST	Management, technology, science	9/74
Volvo Fritid (Sweden)	SCST	Scientific and technical co-operation	6/75
NCR (US)	SCST	Automation technology for retail stores, restaurants	10/76
Rank Taylor Hobson (UK)	Gosstandort	Research into surface measurement and evaluation	5/77
Seibu (Japan)	Soviet government authorities	Joint research into retailing techniques	5/77
Shipping and Shipbuilding			
Svensk Varvs Industri Föreningen (Sweden)	SCST	Modernisation of equipment and engines	5/76
Wärtsila (Finland)	Ministry of Shipbuilding Industry	Scientific technical co-operation; increased trade in ships and ship equipment	1/77
Bos Kalis Westminster Group (Netherlands)	SCST; Ministry of Maritime Fleet; Ministry of River Fleet	Study of underwater formations; development of dredging methods	3/77
Wärtsila (Finland)	Ministry of Shipubilding Industry; SCST	Design of ship structures; standardization of ship engines	7/77

Table A-11 *(contd').*

Western Partner	Soviet Partner	Purpose of Agreement	Announced
Telecommunications			
ITT Corp. (US)	SCST	Telecommunications, electronic components, consumer goods, scientific and technical information publication	7/73
		US-USSR satellite communication link	12/73
Textiles			
Agache (France)	Techmashexport	Textiles and fashion goods	5/74
Karl Mayer (Germany)	SCST	Design and manufacture of equipment for textile industry	1/75
Picanol (Belgium)	SCST	Looms	2/73
MacIntosh Confectie (Netherlands)	Soviet government authorities	Clothing factory; commercial co-operation	1/77
Teijin (Japan)	SCST	Production and marketing of synthetic fibers	7/77
Wood and Product			
International Paper (US)	SCST	Scientific and technical co-operation in pulp and paper industry	1/75
Nihon Chip Boeki, head of 30 Japanese-firm consortium		Woodworking equipment against Soviet wood chips	8/72
Valmet (Finland)	Ministry of Pulp and Paper Industry	Paper-making machinery	9/77
Canadian Institute of the Pulp and Paper Industry	Soviet Pulp and Paper Scientific and Technical Information Institute	New methods of pulp cooking; transportation and storage of sawed timber	10/77
Miscellaneous Equipment			
Nummela (Finland)	SCST	Joint development of fire extinguishers for trucks	8/77

* SCST : State Committee for Science and Technology.
** Renewal.
Source: Eugene Zaleski and Helgard Wienert, *Technology Transfer between East and West* Paris: Organisation for Economic Co-operation and Development, 1980, Table A-31, pp. 349-362.

Table A-12. **Value of Soviet imports, by commodity group, 1970 and 1975-1981**[a]
Millions of US dollars

	1970	1975	1976	1977	1978	1979	1980	1981
Total imports	11 720	37 070	38 212	40 926	50 798	57 958	68 473	73 158
Machinery and equipment	4 166	12 574	13 868	15 587	21 354	22 019	23 198	22 107
Transportation equipment	1 232	3 326	3 123	3 142	3 943	4 335	4 674	4 660
Fuels, lubricants, and related materials	228	1 447	1 374	1 473	b	b	b	269
Coal and coke	124	499	480	b	b	b	b	b
Petroleum and petroleum products	82	695	661	b	b	b	b	b
Ores and concentrates	303	583	336	574	578	747	n.d.	929
Base metals and manufactures	691	3 595	3 304	2 480	3 414	4 499	4 654	4 594
Ferrous metals	593	3 346	3 029	2 466	3 400	4 478	4 635	4 574
Non-ferrous metals	98	249	275	15	14	21	19	20
Chemicals	618	1 747	1 668	1 850	2 188	2 803	3 732	4 031
Rubber and rubber products	192	307	357	n.d.	n.d.	288[c]	320[c]	298
Wood and wood products	248	796	670	723	747	847	1 369	1 305
Textile raw materials and semi-manufactures	561	887	894	1 037	1 025	1 122	1 493	1 271
Cotton fiber	249	271	234	226	93	158	96	46
Wool fiber	120	265	304	373	420	487	502	534
Consumer goods	3 789	12 720	12 574	12 802	14 750	17 834	23 334	27 532
Food	1 590	7 844	7 736	7 523	8 822	11 395	15 156	18 371
Other consumer goods	2 199	4 876	4 838	5 279	5 928	6 440	8 179	9 161
Other	926	2 414	2 967	4 350	6 742	7 799	10 373	10 822

a) Official Soviet statistics using US dollar exchange rates for the Soviet foreign trade (valuta) ruble as announced by the State Bank of the USSR. Imports are f.o.b. Components may not add to totals because of rounding.
b) Not available.
c) Only natural rubber reported.
Source: US Central Intelligence Agency, *Handbook of Economic Statistics, 1983* (CPAS 83-10006; Washington, D.C., September 1983), Table 65, p. 95.

Tableau A-13. **Value of Soviet exports, by commodity group, 1970 and 1975-1981**[a]

Millions of US dollars

	1970	1975	1976	1977	1978	1979	1980	1981
Total exports	12 787	33 407	37 269	45 227	52 435	64 912	76 437	79 377
Machinery and equipment	2 753	6 241	7 219	8 493	10 277	11 378	12 081	10 862
Fuels, lubricants, and related materials	1 986	10 473	12 773	15 853	18 695	27 379	35 874	39 849
Coal and coke[b]	408	1 399	1 350	1 422	418	1 504	1 695	1 597
Petroleum and petroleum products	1 469	8 212	10 210	12 778	14 776	22 211	27 851	30 029
Ores and concentrates	403	868	839	829	887	919	993	1 016
Iron ore	325	692	629	599	628	656	651	656
Base metals and manufactures	1 978	3 813	3 873	2 817	2 987	3 113	3 551	3 511
Ferrous metals	1 351	2 673	2 734	2 814	2 987	3 113	3 551	3 511
Non-ferrous metals	627	1 140	1 138	3	*c*	*c*	*c*	*d*
Chemicals	364	1 027	939	1 063	1 284	1 606	2 139	2 380
Wood and wood products	831	1 916	1 995	2 299	2 339	2 666	3 093	2 632
Lumber	333	780	817	934	936	1 031	1 207	990
Textile raw materials and semi-manufactures	437	976	1 067	1 448	1 313	1 304	1 435	1 545
Cotton fiber	372	917	1 013	1 372	1 229	1 221	1 357	1 472
Consumer goods	1 356	2 377	2 020	2 336	2 311	2 855	2 727	2 759
Food	1 019	1 556	1 106	1 391	1 165	1 661	1 426	1 587
Other consumer goods	337	821	913	945	1 146	1 195	1 301	1 172
Other	2 679	5 716	6 544	10 089	12 342	13 692	14 544	14 823

a) Official Soviet statistics using US dollar exchange rates for the Soviet foreign trade (valuta) ruble as announced by the State Bank of the USSR. Exports are f.o.b. Components may not add to totals because of rounding.
b) Including small amounts of charcoal and other solid fuels.
c) Not available.
d) Negligible.
Source: US Central Intelligence Agency, *Handbook of Economic Statistics, 1983* (CPAS 83-10006 ; Washington, D.C., September 1983), Table 64, p. 94.

Table A-14. **Value of Soviet foreign trade, by country group, 1970, and 1975-1982**[a]

Millions of US dollars

Year		Total trade	Communist countries					Non-Communist countries		
			Total	Eastern Europe	China	Other Asian	Other[b]	Total	Developed countries	Less developed countries
1970	Exports	12 787	8 359	6 752	25	415	1 167	4 428	2 453	1 975
	Imports	11 720	7 630	6 627	22	156	826	4 089	2 814	1 275
1975	Exports	33 407	20 271	16 494	129	480	3 168	13 136	8 588	4 548
	Imports	37 070	19 415	15 723	150	277	3 265	17 655	13 566	4 089
1976	Exports	37 269	21 890	17 432	239	565	3 654	15 378	10 269	5 109
	Imports	38 212	20 088	16 261	179	242	3 406	18 123	14 360	3 763
1977	Exports	45 227	26 009	20 762	161	628	4 458	19 219	12 226	6 993
	Imports	40 926	23 354	18 839	177	400	3 938	17 573	13 416	4 157
1978	Exports	52 435	31 244	24 910	241	726	5 367	21 190	12 920	8 270
	Imports	50 798	30 494	24 661	257	521	5 055	20 304	16 244	4 060
1979	Exports	64 912	36 152	28 380	268	1 081	6 423	28 760	19 578	9 182
	Imports	57 958	32 810	26 761	240	618	5 191	25 148	20 405	4 743
1980	Exports	76 437	41 431	32 216	261	1 201	7 752	35 006	24 934	10 072
	Imports	68 473	36 420	29 407	226	681	6 106	32 053	24 386	7 667
1981	Exports	79 377	43 353	33 774	115	1 445	8 019	36 024	24 416	11 608
	Imports	73 158	37 172	29 399	131	582	7 060	35 986	25 356	10 630
1982	Exports	87 170	47 109	36 288	166	1 638	9 017	40 061	26 224	13 837
	Imports	77 848	42 526	33 566	143	788	8 029	35 322	26 204	9 118

a) Official Soviet statistics using US dollar exchange rates for the Soviet foreign trade (valuta) ruble as announced by the State Bank of the USSR. Exports and imports are f.o.b. Components may not add to totals because of rounding.

b) Cuba, Mongolia and Yugoslavia.

Source: US Central Intelligence Agency, *Handbook of Economic Statistics, 1981* (NF HES 81-001 ; Washington, D.C., November 1981), Table 58, p. 83. and *Handbook of Economic Statistics, 1983* (CPAS 83-10006 ; Washington, D.C., September 1983), Table 63, p. 93.

Table A-15. **Value of OECD imports from the USSR, by sector of origin, 1970-1982**[a]

Millions of US dollars

Sector of origin	1970	1971	1972	1973	1974	1975	1976	1977	1978	1979	1980	1981	1982
Agricultural products													
Cereals	39.9	54.7	28.8	3.9	2.3	2.7	0.7	0.0[b]	0.0	2.6	0.1	0.1	0.1
Fruits and vegetables	5.9	14.0	6.7	7.7	12.4	6.4	3.0	4.5	2.3	3.6	5.5	10.1	7.3
Inedible crude materials	50.4	49.6	42.1	58.2	65.1	58.6	86.4	78.2	90.9	118.9	116.6	118.9	90.3
Wood and lumber	278.0	271.5	288.9	523.2	710.0	606.4	597.1	714.1	674.5	878.7	904.4	643.9	564.7
Textile fibers	57.4	84.6	157.5	235.1	362.7	361.2	441.1	533.1	411.0	395.5	392.9	526.9	454.7
Live animals	12.6	12.2	14.1	28.0	48.8	42.3	46.5	46.2	55.0	78.6	65.9	70.2	76.2
Tobacco (unmanfactured)	0.8	1.0	0.4	0.7	0.5	2.3	2.7	2.4	1.9	1.8	1.0	3.3	1.4
Other products	37.8	44.7	43.4	61.9	167.3	211.0	104.6	79.9	47.3	39.6	83.3	91.5	74.1
Total	482.9	532.2	581.8	918.6	1369.0	1290.9	1282.1	1458.4	1282.8	1519.3	1569.7	1464.8	1268.8
Industrial products	2071.0	2310.5	2579.0	3970.1	6550.5	7202.9	9483.4	10669.8	11812.5	17048.4	22218.9	22302.1	23791.1
Grand total	2553.9	2842.7	3160.9	4888.7	7919.6	8493.7	10765.5	12128.2	13095.4	18567.7	23788.5	23766.9	25059.9

a) Components may not add to toals because of rounding.
b) Less than 0.1 million US dollars.
Source: East-West Technology Transfer Data Base, OECD, Paris.

Table A-16. **Deflated value of OECD imports from the USSR, by sector of origin, 1970-1981**[a]

Millions of 1970 US dollars

Sector of origin	1970	1971	1972	1973	1974	1975	1976	1977	1978	1979	1980	1981
Agricultural products												
Cereals	33.9	49.0	26.8	2.2	1.0	0.9	0.3	0.0[b]	0.0	0.7	0.0	0.0
Fruits and vegetables	5.9	11.5	5.6	6.1	8.6	3.7	1.9	3.2	1.1	2.1	3.0	4.7
Inedible crude materials	50.4	51.1	26.2	25.0	29.0	34.1	36.5	42.7	38.9	36.2	35.5	55.1
Wood and lumber	278.0	249.1	267.9	337.5	274.5	273.0	280.2	271.7	262.1	304.3	241.4	202.5
Textile fibers	57.4	78.3	123.8	172.6	156.1	185.3	232.7	212.3	183.2	151.8	142.1	181.3
Live animals	12.6	12.1	13.4	19.8	30.4	26.0	26.0	25.0	24.0	28.9	23.8	22.0
Tobacco (unmanufactured)	0.8	1.0	0.5	0.5	0.3	1.2	1.2	1.5	1.0	1.0	0.5	1.7
Other products	37.8	46.8	44.6	58.6	131.6	124.3	70.4	51.1	31.2	22.9	34.7	41.9
Total	482.9	498.9	508.9	622.3	631.5	648.6	649.1	607.6	541.6	547.8	481.0	509.1
Industrial products	2071.0	2208.1	2667.6	3141.7	2987.7	2914.9	3991.5	4334.3	4133.0	3900.6	4017.5	3807.3
Grand total	2553.9	2707.0	3176.4	3764.0	3619.3	3563.4	4640.6	4941.9	4674.6	4448.5	4498.5	4316.5

a) Components may not add to toals because of rounding.
b) Less than 0.1 million US dollars.
Source: East-West Technology Transfer Data Base, OECD, Paris.

Table A-17. **Value of OECD industrial goods imports from the USSR, by branch of origin, 1970-1982[a]**

Millions of US dollars

Branch of origin	1970	1971	1972	1973	1974	1975	1976	1977	1978	1979	1980	1981	1982
1. Products of mining													
Crude minerals	39.5	40.4	39.9	45.3	60.1	81.0	86.2	90.0	69.2	56.9	76.6	64.4	46.9
Metalliferrous ores	90.0	95.1	83.9	96.2	117.8	191.0	177.6	115.8	109.3	75.4	48.7	37.2	38.8
Coal	143.0	162.8	156.2	170.8	311.8	462.2	451.4	427.6	380.2	411.0	450.7	259.9	216.7
Petroleum and gas	314.6	466.9	422.9	546.6	1092.3	1464.1	2661.9	3188.8	3558.8	4760.5	7957.7	9213.2	9781.0
Total	587.1	765.2	703.0	858.9	1582.1	2198.3	3377.0	3822.2	4117.4	5303.8	8533.6	9574.7	10083.4
2. Coke and manufactured gas	34.6	33.9	32.7	39.8	44.9	71.4	64.0	67.9	74.0	72.9	104.8	91.7	84.8
3. Energy	1.4	2.1	3.3	2.8	7.5	5.7	6.0	9.9	10.5	12.0	36.0	33.5	67.0
4. Food processing													
Products of milling industry	0.2	0.1	0.2	0.2	0.4	0.4	0.4	0.2	0.1	0.0[b]	0.1	0.1	0.2
Other food processing; beverage industry	3.2	4.0	5.9	9.0	8.8	11.0	13.8	16.8	23.2	35.9	35.8	23.2	28.0
Tobacco manufactures	0.0	0.0	0.0	0.0	0.0	0.1	0.0	0.0	0.0	0.0	0.0	0.1	0.0
Meat	5.0	5.9	10.1	18.5	19.0	20.9	29.1	44.1	40.2	39.7	38.8	33.6	29.7
Dairy products	1.2	1.6	1.0	0.6	1.7	0.9	0.3	0.3	0.2	0.1	0.1	0.1	0.1
Other processed food	114.5	137.6	108.4	119.6	258.6	231.1	143.7	132.4	145.8	136.3	136.0	129.4	111.2
Total	124.1	149.2	125.6	147.9	288.4	264.3	187.2	193.7	209.5	212.0	210.8	186.4	169.1
5. Textiles													
Knitted products	0.1	0.0	0.3	0.2	0.4	0.3	0.2	0.1	0.2	0.3	0.5	0.5	0.2
Other textiles	6.2	5.0	5.2	6.4	9.9	8.8	22.7	24.4	28.1	34.6	21.9	12.9	15.0
Manufactured textiles	15.0	17.5	27.3	41.6	41.0	42.4	55.7	67.0	71.8	85.2	65.5	51.3	44.0
Total	21.3	22.6	32.7	48.3	51.2	51.5	78.7	91.4	100.1	120.1	87.9	64.8	59.2
6. Clothing	0.7	0.5	0.9	1.1	1.3	0.9	1.2	1.3	1.4	1.7	1.8	2.0	1.8
7. Leather, shoes, and furs	8.6	7.8	9.5	11.6	15.6	15.7	20.1	19.4	19.5	22.1	23.5	18.6	11.2
8. Paper, pulp, and processed wood													
Processed wood except furniture	217.5	209.9	243.2	359.3	488.6	360.3	513.5	564.5	534.9	662.1	777.6	490.4	475.4
Furniture	0.1	0.1	0.2	0.5	0.9	1.6	2.2	4.0	8.3	9.0	11.9	13.2	13.7
Paper and paper products	24.1	19.5	26.4	41.3	67.3	44.3	76.8	93.0	119.6	136.9	172.1	171.9	122.1
Total	241.7	229.5	269.8	401.1	556.8	406.2	592.5	661.4	662.8	807.9	961.6	675.5	611.2
9. Products of printing industry	0.6	0.8	1.0	1.4	1.3	2.2	2.6	2.7	5.5	6.2	5.0	4.8	5.3

Table A-17 *(cont'd).*

Branch of origin	1970	1971	1972	1973	1974	1975	1976	1977	1978	1979	1980	1981	1982
10. Glass and china													
Glass and glass products	2.8	2.9	3.9	5.5	4.5	3.8	8.7	10.5	10.1	10.6	9.7	8.8	5.4
China	0.3	0.2	0.2	0.4	0.6	0.8	0.9	1.1	1.1	1.9	1.4	1.5	1.4
Total	3.1	3.1	4.2	6.0	5.1	4.6	9.6	11.5	11.2	12.5	11.1	10.3	6.8
11. Chemicals													
Basic industrial chemicals	44.2	46.1	49.6	81.7	199.3	175.8	281.1	575.1	740.5	1246.1	1034.3	813.7	668.6
Fertilizers and pesticides	21.8	25.0	32.3	53.0	83.8	105.2	67.7	73.6	75.4	87.7	155.1	110.9	117.0
Synthetic resins, plastic and man-made fibers	1.3	2.0	2.0	5.6	8.5	8.7	19.0	25.9	30.0	46.4	64.2	80.1	106.0
Paints and varnishes	0.1	0.3	0.2	0.3	0.3	0.2	0.2	0.0	0.0	0.1	0.1	0.0	0.0
Drugs, medicines	2.3	4.0	3.4	3.0	6.1	7.0	4.5	5.7	5.5	7.0	5.7	8.4	8.6
Soaps, perfumes	0.2	0.2	0.1	0.1	0.2	0.3	0.3	0.2	0.3	0.6	0.2	0.1	0.0
Gasoline, oils, other petroleum products	311.9	402.1	427.7	825.7	1842.2	2244.6	2502.5	2698.5	3166.2	5520.7	6403.4	6861.2	8241.5
Other products of petroleum and coal	...	...	...	0.0	0.0	...	...	0.0	0.0	0.0	0.0	0.0	0.0
Rubber products	0.3	0.5	0.5	0.5	0.8	1.1	1.3	1.1	1.5	2.4	3.3	2.2	2.6
Other chemicals	3.6	5.1	5.1	5.7	8.6	9.6	8.2	10.0	11.7	11.2	12.2	11.1	9.2
Total	385.7	485.3	521.0	975.5	2149.8	2552.4	2884.9	3390.1	4031.2	6922.1	7678.6	7887.7	9153.4
12. Other non-metallic mineral products	4.1	2.3	7.2	9.0	4.0	5.6	4.7	4.3	2.3	4.0	4.9	3.3	3.7
13. Metallurgy													
Iron and steel	98.1	94.3	110.7	129.3	115.1	105.5	109.1	116.1	93.5	124.8	162.7	92.1	154.7
Non-ferrous metals	301.0	260.0	372.4	707.6	879.1	478.7	474.9	510.6	585.6	769.6	852.8	600.6	639.7
Total	399.1	354.3	483.1	836.9	994.2	584.2	584.0	626.7	679.1	894.4	1015.5	692.8	794.4
14. Machinery, equipment, and metal products													
Engines and turbines	0.2	0.7	0.2	6.8	1.3	6.4	11.3	9.5	9.9	11.0	8.2	1.2	1.3
Agricultural machinery	3.4	2.9	6.4	10.7	14.9	23.7	27.9	21.5	17.0	27.8	26.0	17.4	13.5
Metal and woodworking machinery	19.8	30.0	35.1	36.3	39.2	63.4	57.2	47.2	37.9	43.3	58.4	44.8	34.4
Textile machinery	0.6	0.4	0.7	2.4	2.0	5.1	1.3	1.1	1.9	3.8	4.2	3.5	1.8
Paper and pulp machinery	0.2	4.9	0.2	0.1	0.2	1.3	5.2	1.3	1.1	0.6	2.3	2.6	0.6
Printing and bookbinding machinery	1.4	2.0	0.7	1.1	0.3	0.9	1.0	0.6	0.6	0.5	0.3	0.5	0.2
Food processing machines	0.3	0.3	0.2	0.3	1.1	0.9	0.8	0.4	0.4	0.4	0.6	0.6	0.8
Construction and mining machinery	3.3	4.7	2.1	4.3	4.8	7.5	9.9	5.7	5.1	4.5	5.9	15.0	7.4
Glass working machinery	0.3	0.0	0.0	0.0	0.0	0.0	0.0	0.0	0.0	0.0	0.0	0.0	0.0
Other machinery for special industries	0.6	1.5	12.0	42.4	21.3	13.6	19.5	3.3	4.0	1.1	2.5	3.8	3.2
Office and computing machinery	0.5	0.4	1.5	0.7	0.8	1.5	2.6	2.3	3.0	1.7	2.7	2.2	3.3

Electrical appliances and housewares	0.1	0.1	0.2	0.3	0.2	0.4	0.3	0.4	0.5	0.7	0.7	0.9	0.7
Ships	21.0	35.7	19.0	44.2	23.0	46.8	83.8	26.1	55.4	212.9	95.5	101.0	20.6
Railroad equipment	1.6	1.8	2.6	4.5	4.8	11.2	6.8	11.8	3.2	2.8	1.8	3.6	3.2
Motor vehicles	8.7	13.3	29.8	57.6	61.6	112.7	142.3	160.4	204.7	314.1	251.0	203.8	229.7
Motorcycles and cycles	0.4	0.6	0.7	1.0	2.6	2.8	2.1	0.9	0.7	0.6	1.6	0.7	0.3
Aircraft	1.0	0.8	9.8	4.1	5.6	0.7	0.9	1.1	12.0	5.4	74.6	9.1	8.3
Scientific, measuring and control equipment	2.1	1.8	3.3	7.7	5.9	7.0	5.7	6.0	7.1	7.3	6.5	9.6	6.9
Photographic and optical goods	4.1	5.5	6.9	9.8	11.4	13.7	12.4	11.4	14.8	17.5	16.0	9.6	6.0
Watches and clocks	2.3	2.9	4.4	6.9	8.6	9.4	9.5	12.3	11.3	14.1	19.5	11.5	10.3
Structural metal products	7.7	7.6	13.0	22.2	14.9	20.4	30.3	12.6	29.3	8.6	19.4	23.0	27.9
Cutlery and hand tools	1.3	1.5	1.9	2.9	2.8	3.3	3.5	3.5	4.9	4.5	6.2	6.1	4.5
Other electrical apparatus	1.1	0.6	8.7	6.5	5.7	2.6	3.1	1.7	1.4	2.2	2.3	2.0	1.4
Total	90.4	128.7	178.1	293.6	259.0	388.1	474.5	376.4	465.0	728.7	660.9	510.7	421.1
15. Other industrial commodities	168.5	125.2	207.1	336.3	589.2	651.7	1196.5	1390.6	1423.0	1927.9	2882.9	2545.4	2318.5
16. Total industrial imports	2071.0	2310.5	2579.0	3970.1	6550.5	7202.9	9483.4	10669.8	11812.5	17048.4	22218.9	22302.1	23791.1

a) The classification corresponds to the USSR branch of origin classification. Components may not add to totals because of rounding.
b) Less than 0.1 million US dollars.
Source: East-West Technology Transfer Data Base, OECD, Paris.

Table A-18. **Deflated value of OECD industrial goods imports from the USSR, by branch of origin, 1970-1981**[a]

Millions of 1970 US dollars

Branch of origin	1970	1971	1972	1973	1974	1975	1976	1977	1978	1979	1980	1981
1. Products of mining												
Crude minerals	39.5	42.1	41.5	43.2	47.3	47.6	57.6	59.0	40.5	31.7	28.2	27.3
Metalliferrous ores	90.0	99.6	99.4	79.1	64.6	114.0	118.1	78.6	70.2	35.4	20.4	19.4
Coal	143.0	125.0	122.6	126.5	155.7	142.1	150.3	142.3	118.5	118.2	105.9	54.5
Petroleum and gas	314.6	315.6	323.1	297.7	239.0	301.0	533.1	593.5	601.7	500.2	565.2	634.1
Total	587.1	582.4	586.5	546.4	506.6	604.7	859.0	873.4	830.9	685.5	719.8	735.3
2. Coke and manufactured gas	34.6	26.0	25.6	29.5	22.4	22.0	21.3	22.6	23.1	21.0	24.6	19.2
3. Energy	1.4	2.1	2.5	2.0	3.6	2.7	2.9	3.0	2.9	3.0	7.0	7.1
4. Food processing												
Products of milling industry	0.2	0.1	0.2	0.1	0.2	0.1	0.1	0.1	0.0[b]	0.0	0.0	0.0
Other food processing; beverage industry	3.2	4.2	5.4	6.7	5.7	7.7	9.0	13.5	14.3	17.7	23.6	23.2
Tobacco manufactures	0.0	0.0	0.0	0.0	0.0	0.1	0.0	0.0	0.0	0.0	0.0	0.0
Meat	5.0	5.2	7.8	11.0	10.1	9.2	9.2	9.6	6.8	4.9	4.6	3.3
Dairy products	1.2	1.1	0.3	0.2	0.3	0.2	0.1	0.1	0.0	0.0	0.0	0.0
Other processed foods	114.5	123.0	103.1	102.6	128.6	146.4	98.2	87.5	80.7	63.2	59.0	55.9
Total	124.1	133.6	116.6	120.5	144.9	163.8	116.6	110.7	101.8	85.8	87.2	82.6
5. Textiles												
Knitted products	0.1	0.0	0.2	0.1	0.2	0.2	0.1	0.0	0.1	0.1	0.2	0.2
Other textiles	6.2	4.7	4.1	4.7	4.3	4.5	11.6	9.5	12.4	13.1	7.9	4.4
Manufactured textiles	15.0	16.6	23.0	25.9	18.3	19.7	25.6	26.3	24.6	28.1	19.2	17.4
Total	21.3	21.3	27.3	30.7	22.7	24.4	37.3	35.8	37.0	41.3	27.3	22.0
6. Clothing	0.7	0.6	0.6	0.5	0.7	0.5	0.6	0.6	0.6	0.6	0.6	0.8
7. Leather, shoes, and furs	8.6	7.2	8.5	8.8	11.6	12.4	11.3	11.3	10.9	12.5	9.9	8.4
8. Paper, pulp, and processed wood												
Processed wood except furniture	217.5	197.9	227.9	244.7	211.4	171.8	245.9	233.4	213.8	233.8	212.8	153.3
Furniture	0.1	0.1	0.2	0.3	0.5	0.7	0.9	1.5	2.6	2.4	3.0	4.0
Paper and paper products	24.1	18.0	24.8	33.0	32.2	16.9	33.5	39.0	57.6	55.6	58.6	60.6
Total	241.7	216.1	252.9	278.1	244.0	189.4	280.4	263.9	274.0	291.8	274.4	217.9
9. Products of printing industry	0.6	0.7	0.8	0.9	0.8	1.1	1.2	1.2	2.2	2.2	1.7	1.9
10. Glass and china												
Glass and china products	2.8	4.5	4.5	3.6	2.9	2.5	12.1	19.7	9.1	4.6	3.9	3.9
China	0.3	0.3	0.3	0.3	0.4	0.5	1.3	2.0	1.0	0.8	0.6	0.7
Total	3.1	4.8	4.7	3.9	3.3	3.0	13.4	21.7	10.1	5.4	4.5	4.5
11. Chemicals												
Basic industrial chemicals	44.2	53.6	61.9	82.0	90.2	109.3	172.9	343.6	397.1	401.6	302.7	284.2

Drugs, medicines	2.3	2.8	11.3	1.8	2.8	3.3	2.1	1.1	2.9	1.8	1.8	17.1
Soaps, perfumes	0.2	0.2	0.1	0.1	0.2	0.2	0.2	0.1	0.1	0.2	0.1	0.0
Gasoline, oils, other petroleum products	311.9	307.8	330.2	445.3	396.6	454.6	487.8	489.8	517.1	533.8	442.7	455.3
Other products of petroleum and coal	...	...	...	0.0	0.0	...	...	0.0	0.0	0.0	0.0	0.0
Rubber products	0.3	0.3	0.4	0.4	0.5	0.4	1.0	0.5	1.0	1.8	2.4	1.5
Other chemicals	3.6	3.8	5.5	7.2	7.1	6.9	5.7	6.1	5.5	4.1	4.3	4.1
Total	385.7	393.9	431.6	587.6	545.4	619.4	712.2	896.6	979.2	1001.0	823.0	827.2
12. Other non-metallic mineral products	4.1	3.5	8.2	5.9	2.6	3.6	6.5	8.1	1.8	1.9	1.9	1.4
13. Metallurgy												
Iron and steel	98.1	102.9	128.7	111.9	69.9	52.9	68.9	78.0	60.9	53.5	61.5	40.0
Non-ferrous metals	301.0	384.5	696.4	971.6	907.9	614.5	588.7	532.2	535.7	565.0	506.6	401.8
Total	399.1	487.4	825.1	1083.5	977.8	667.4	657.6	610.2	596.6	618.5	568.2	441.8
14. Machinery, equipment, and metal products												
Engines and turbines	0.2	0.1	0.2	4.1	0.7	3.0	5.1	3.9	3.3	3.2	2.3	0.4
Agricultural machinery	3.4	2.6	5.0	6.8	8.5	11.7	13.0	8.5	5.5	7.9	7.1	5.6
Metal and woodworking machinery	19.8	27.4	29.6	25.2	21.2	28.2	24.4	18.5	12.1	12.4	15.8	13.8
Textile machinery	0.6	0.3	0.6	1.8	1.3	3.0	0.7	0.9	0.8	1.4	1.5	1.3
Paper and pulp machinery	0.2	4.3	0.2	0.0	0.1	0.8	2.8	1.1	0.5	0.2	0.8	0.9
Printing and bookbinding machinery	1.4	1.7	0.6	0.8	0.2	0.5	0.5	0.5	0.2	0.2	0.1	0.2
Food processing machines	0.3	0.3	0.2	0.2	0.7	0.6	0.5	0.4	0.2	0.2	0.2	0.2
Construction and mining machinery	3.3	4.3	1.7	3.1	3.0	4.1	5.1	2.6	2.0	1.5	1.8	5.1
Glass working machinery	0.3	0.0	0.0	0.0	0.0	0.0	0.0	0.0	0.0	0.0	0.0	0.0
Other machinery for special industries	0.6	1.4	9.4	27.2	12.1	6.7	9.1	1.3	1.3	0.3	0.7	1.2
Office and computing machinery	0.5	0.3	1.5	0.6	0.7	1.2	2.1	1.8	2.1	1.1	1.7	1.7
Electrical industrial machinery and apparatus	3.5	3.7	6.6	5.5	5.2	8.6	11.7	6.8	5.7	5.6	13.4	8.8
Radio, TV, and communication equipment	5.0	4.6	11.4	10.2	13.2	12.1	12.1	14.0	14.2	14.4	11.0	11.1
Electrical appliances and housewares	0.1	0.1	0.1	0.2	0.2	0.2	0.2	0.2	0.3	0.3	0.3	0.5
Ships	21.0	32.5	15.4	30.6	14.6	25.2	43.6	13.2	23.0	99.6	35.4	36.8
Railroad equipment	1.6	1.6	2.1	3.2	3.0	5.9	3.4	5.5	1.3	1.0	0.6	1.5
Motor vehicles	8.7	11.8	24.2	40.2	38.5	59.2	71.9	75.7	82.6	113.0	85.2	82.4
Motorcycles and cycles	0.4	0.5	0.6	0.7	1.6	1.5	1.1	0.4	0.3	0.2	0.5	0.3
Aircraft	1.0	0.7	7.9	2.8	3.7	0.4	0.4	0.5	5.1	2.3	27.4	3.3
Scientific, measuring and control equipment	2.1	1.7	3.0	5.6	4.1	4.3	3.5	3.3	3.3	3.2	2.7	4.7
Photographic and optical goods	4.1	5.1	5.7	6.5	7.6	8.0	7.0	6.0	6.6	7.2	6.3	4.4
Watches and clocks	2.3	2.7	3.6	4.6	5.7	5.5	5.4	6.5	5.0	5.8	7.7	5.3
Structural metal products	7.7	6.3	9.9	15.2	10.4	11.3	15.8	5.6	10.4	2.6	5.7	7.9
Cutlery and hand tools	1.3	1.2	1.4	2.0	2.1	1.9	2.1	1.9	2.4	1.7	2.4	2.4
Other electrical apparatus	1.1	0.6	8.4	4.8	4.0	1.7	2.0	1.0	0.7	1.0	1.0	1.0
Total	90.4	116.0	149.5	202.0	162.5	205.5	243.5	180.3	188.9	286.3	231.8	200.8
15. Other industrial commodities	168.5	212.6	227.1	241.4	338.8	395.1	1027.7	1294.9	1073.0	843.9	1235.6	1236.3
16. Total industrial imports	2071.0	2208.1	2667.6	3141.7	2987.7	2914.9	3991.5	4334.3	4133.0	3900.6	4017.5	3807.3

a) The classification corresponds to the USSR branch of origin classification. Components may not add to totals because of rounding.
b) Less than 0.1 million US dollars.
Source: East-West Technology Transfer Data Base, OECD, Paris.

Table A-19. **Value of OECD capital goods imports from the USSR, by type of product, 1970-1982**[a]

Millions of US dollars

Type of product	1970	1971	1972	1973	1974	1975	1976	1977	1978	1979	1980	1981	1982
1. Stationary power plants and water engineering													
Stationary power plants	5.4	6.8	9.9	12.0	9.8	19.7	39.7	24.1	36.2	19.7	50.4	33.5	39.2
Water turbines and engines	0.0[b]	0.1	0.2	1.0	0.3	1.8	6.6	5.4	9.2	9.5	3.8	1.0	1.1
Total	5.4	6.9	10.1	13.0	10.1	21.5	46.3	29.6	45.3	29.2	54.1	34.5	40.3
2. Electric power distribution	1.1	0.9	2.5	6.1	4.9	2.9	2.6	2.3	1.9	2.0	2.3	1.5	1.9
3. Liquid fuel, gas, and water distribution													
Tubes and pipes	2.9	6.5	4.6	7.2	10.3	12.7	17.9	10.4	9.7	9.2	7.9	5.4	7.0
Pumps and centrifuges	2.2	1.4	2.5	2.0	2.2	3.7	5.2	3.3	2.0	3.5	6.6	8.0	7.5
Total	5.1	7.9	7.1	9.2	12.5	16.4	23.1	13.7	11.8	12.7	14.6	13.4	14.5
4. Transport equipment													
Rail transport equipment	0.0	0.4	1.7	3.5	3.9	8.5	4.2	10.5	2.1	0.2	0.1	0.4	0.0
Road transport equipment	1.9	3.1	6.0	7.9	7.6	5.6	7.7	9.2	6.1	17.1	15.1	5.5	7.8
Aircraft	0.6	0.1	8.4	1.8	4.4	0.2	0.0	0.3	8.4	1.1	57.9	5.4	4.6
Sea and river transport equipment	16.4	32.1	15.6	39.1	13.9	36.3	67.4	19.0	36.8	198.6	74.4	83.7	5.3
Total	18.9	35.7	31.7	52.3	29.8	50.6	79.3	39.1	53.5	217.0	147.5	95.0	17.7
5. Agricultural equipment													
Field machinery	0.1	0.0	0.4	0.5	0.4	1.1	1.3	0.8	0.9	1.5	0.8	0.8	0.3
Dairy farm equipment	0.0	0.0	0.0	0.0	0.0	0.0	0.0	0.0	0.0	0.0	0.0	0.0	0.0
Tractors	3.3	2.9	5.9	10.2	14.5	22.6	26.6	20.6	16.1	26.2	25.2	16.6	13.2
Presses for making wine, juices, etc.	0.0	0.0	0.0	0.0	0.0	0.0	0.0	0.0	0.0	0.0	0.0	0.0	0.0
Other agricultural machinery	0.0	0.0	0.0	0.0	0.0	0.0	0.0	0.0	0.0	0.2	0.0	0.1	0.0
Total	3.4	2.9	6.4	10.7	14.9	23.7	27.9	21.5	17.0	27.8	26.0	17.5	13.5
6. Construction and mining machinery	4.3	7.0	10.1	5.1	6.5	10.4	12.5	8.0	7.0	7.2	7.7	17.5	10.1
7. Engineering, welding, and metallurgical equipment (excluding furnaces)													
Machine tools	17.4	25.4	23.4	26.5	29.9	39.9	44.0	35.7	29.0	33.4	47.9	37.5	29.4
Metallurgical machinery	1.2	2.9	9.9	7.9	6.3	18.7	8.2	6.9	5.1	5.5	5.8	3.3	2.1
Welding appliances and equipment	0.0	0.0	0.1	0.0	0.0	0.0	0.1	0.4	0.0	0.4	0.0	0.0	0.0
Total	18.6	28.4	33.3	34.4	36.3	58.6	52.4	43.0	34.1	39.4	53.8	40.8	31.5
8. Industrial and laboratory furnaces and gas generators	0.8	4.1	2.4	2.0	1.2	7.1	0.7	2.2	2.3	1.8	4.0	4.0	4.7
9. Machine and hand tools for working minerals, wood, plastic, etc.	0.2	0.2	0.2	0.4	0.6	0.5	0.6	0.7	0.8	1.2	1.2	1.2	0.4

Telecommunications	2.2	1.2	4.8	3.8	5.5	5.3	7.0	6.1	5.6	7.8	8.1	9.1	10.0
Computers	0.3	0.1	0.9	0.3	0.3	0.8	1.3	0.7	2.2	1.3	1.7	0.9	1.9
Optical instruments	0.4	0.4	0.8	1.0	1.3	1.2	1.1	1.3	1.4	1.3	1.3	1.3	0.9
Other electronics	0.0	0.0	0.0	3.9	0.6	0.0	0.1	0.0	0.4	0.0	0.0	1.4	0.0
Total	2.9	1.7	6.4	9.1	7.6	7.3	9.4	8.1	9.5	10.4	11.1	12.7	12.7
12. Machinery for special industries													
Textile and leather machinery	0.8	0.6	0.8	2.5	2.2	5.7	1.9	1.9	2.4	4.5	4.6	3.9	1.9
Pulp and paper mill machinery	0.2	4.9	0.2	0.1	0.2	1.3	5.2	1.3	1.1	0.6	2.3	2.6	0.6
Food processing machinery	0.4	0.4	0.3	0.4	1.2	1.1	0.9	0.5	0.5	0.7	0.7	0.7	0.9
Glassworking machinery	0.3	0.0	0.0	0.0	0.0	0.0	0.0	0.0	0.0	0.0	0.0	0.0	0.0
Printing and bookbinding equipment	1.4	2.0	0.7	1.1	0.3	0.9	1.0	0.6	0.6	0.5	0.3	0.5	0.2
Total	3.0	7.8	2.1	4.1	3.9	9.0	8.8	4.3	4.6	6.2	7.9	7.6	3.5
13. Mechanical handling equipment and storage tanks													
Mechanical handling equipment	2.4	4.1	7.3	3.3	1.9	4.7	7.0	2.9	4.0	3.7	6.1	3.8	4.2
Storage tanks	2.1	0.1	0.2	1.0	1.5	0.9	0.5	0.1	0.0	0.4	0.1	4.3	1.3
Total	4.5	4.3	7.5	4.3	3.4	5.6	7.5	3.0	4.1	4.1	6.2	8.1	5.6
14. Office machines	0.0	0.0	0.0	0.0	0.0	0.3	0.9	1.2	0.6	0.0	0.4	0.0	0.3
15. Medical apparatus, instruments, and furniture	0.3	0.3	0.8	0.8	0.7	0.9	0.9	1.2	0.6	1.1	2.0	1.1	1.5
16. Heating and cooling of buildings and vehicles	0.0	0.0	0.0	0.1	0.0	0.0	0.1	0.0	0.0	0.1	0.3	0.3	1.0
17. Measuring, controlling, and scientific instruments	1.7	1.4	2.5	3.5	4.7	6.5	4.6	5.2	5.3	6.5	5.2	7.1	6.0
18. Finished structural parts and structures	3.3	4.1	9.0	20.7	9.9	15.3	11.0	2.1	3.1	0.5	0.6	0.3	0.3
19. Other capital equipment													
Other non-electrical machinery	0.7	1.7	12.0	42.5	21.6	13.8	19.6	3.4	4.2	1.2	2.8	3.9	3.3
Other equipment	3.5	4.4	3.6	0.8	1.0	0.7	1.4	2.6	1.4	0.8	1.5	5.5	2.2
Total	4.2	6.2	15.6	43.3	22.6	14.5	21.0	6.0	5.6	2.0	4.3	9.4	5.5
Grand total	77.7	119.9	155.1	220.7	171.6	252.8	312.3	193.2	212.5	373.9	351.6	274.3	173.6

a) Components may not add to totals because of rounding.
b) Less than 0.1 million US dollars.
Source: East-West Technology Transfer Data Base, OECD, Paris.

Table A-20. **Value of OECD imports of technology-based intermediate goods from the USSR, by type of product, 1970-1982**[a]

Millions of US dollars

Type of products	1970	1971	1972	1973	1974	1975	1976	1977	1978	1979	1980	1981	1982
1. Parts and accessories of machinery and equipment	10.0	8.9	10.1	16.0	20.6	20.1	20.9	26.0	28.6	36.6	34.3	31.0	22.4
2. Parts and accessories of transport equipment	6.3	6.3	7.9	10.5	16.7	20.3	25.7	17.7	33.1	33.3	53.6	36.2	36.2
3. Paper manufactures	7.5	9.5	8.8	18.1	33.5	20.2	34.3	32.2	41.1	48.0	62.6	53.0	33.5
4. Textile manufactures	1.3	1.4	2.1	2.6	3.1	2.6	2.7	2.7	4.2	4.3	7.1	4.1	4.5
5. Synthetic rubber	0.7	0.8	0.9	3.3	5.1	4.4	8.6	10.4	12.0	13.7	16.8	20.8	27.3
6. Synthetic fibers	0.4	0.3	0.3	0.6	2.2	1.5	2.1	3.1	4.5	6.3	6.5	4.5	2.7
7. Special manufactures of leather, rubber, wood, glass, and minerals	0.1	0.2	0.4	0.4	0.9	0.5	0.8	2.0	1.6	3.0	4.0	3.9	3.4
8. Miscellaneous mineral manufactures	0.3	0.1	0.7	0.1	0.1	0.0[b]	0.2	0.0	0.1	0.1	0.0	0.0	0.1
9. Iron and steel	95.2	87.6	105.6	122.0	104.7	92.7	91.0	105.4	83.7	115.4	154.8	86.7	147.7
10. Non-ferrous metals	278.8	238.8	350.9	675.1	832.2	426.4	420.9	441.3	508.9	700.5	782.6	547.3	600.7
11. Organic and inorganic chemical elements	47.2	47.3	51.6	87.8	213.2	185.7	289.8	606.0	783.2	1355.7	1221.8	1018.8	825.6
12. Final chemical manufactures	0.4	0.8	0.9	1.8	1.7	1.2	2.2	1.1	0.8	2.0	1.9	1.2	1.3
13. Chemical and plastic materials	28.8	34.2	41.0	58.4	99.1	123.5	86.8	101.3	118.0	139.3	224.8	190.7	213.3
14. Other intermediate goods	0.8	0.7	0.7	1.8	1.9	1.7	1.7	1.9	2.4	3.0	3.1	5.6	7.6
Total	477.9	436.7	581.9	998.6	1335.0	900.8	987.8	1351.1	1622.1	2461.1	2573.9	2003.9	1926.0

a) Components may not add to totals because of rounding.
b) Less than 0.1 million US dollars.
Source: East-West Technology Transfer Data Base, OECD, Paris.

Table A-21. **Value of Soviet imports of capital goods from Eastern Europe, by end-use branch, 1970, 1973 and 1975-1980[a]**

Millions of US dollars

End-use branch	1970	1973	1975	1976	1977	1978	1979	1980
Electricity	178.09	305.39	402.67	503.64	624.07	810.57	954.65	1045.20
Fuel and mining	48.24	62.55	121.57	101.95	157.17	304.65	421.77	451.32
Metallurgy	116.03	85.31	93.60	93.75	176.94	241.40	220.93	335.37
Engineering and metalworking	156.20	283.16	326.28	383.52	494.66	572.61	678.02	729.84
Chemicals	140.94	254.30	358.92	415.62	481.11	611.60	656.98	670.12
Building materials	23.32	39.36	73.57	82.98	82.44	93.03	109.21	123.23
Wood and paper	20.54	22.60	35.31	28.77	49.30	73.81	78.99	73.00
Textile	0.00[b]	0.00	0.00	0.00	0.00	0.00	0.00	0.00
Other light industry	105.57	182.83	269.58	302.85	388.28	527.91	613.94	659.85
Food and food processing	115.64	160.95	201.63	177.96	225.12	291.06	388.27	443.70
Total industry	904.57	1396.45	1883.12	2091.04	2679.09	3526.64	4122.70	4531.64
Construction	49.76	57.45	40.10	39.55	117.92	133.81	146.77	169.63
Agriculture, including tractors	165.87	407.24	441.94	592.55	718.46	860.15	979.97	1118.55
Transport and communication	1125.97	1659.65	2699.93	2521.15	2596.03	3275.00	3822.82	4073.80
Trade	0.00	0.00	0.00	0.00	0.00	0.00	0.00	0.00
Other	504.21	1001.67	1449.20	1306.74	1687.80	1834.62	2353.74	2536.40
Grand total	2750.36	4522.46	6514.28	6551.03	7699.30	9630.22	11426.08	12430.01

a) Components may not add to totals because of rounding.
Gb) Less than 0,01 million US dollars.
Source: East-West Technology Transfer Data Base, OECD, Paris.

Table A-22. **Value of Soviet exports of capital goods to Eastern Europe by end-use branch, 1970, 1973 and 1975-1980**[a]

Millions of US dollars

End-use branch	1970	1973	1975	1976	1977	1978	1979	1980
Electricity	38.09	268.07	253.44	300.24	419.27	469.81	544.54	644.98
Fuel and mining	27.05	129.80	168.90	214.92	235.11	250.44	257.90	249.41
Metallurgy	13.97	70.36	101.49	102.75	90.32	165.27	194.42	275.93
Engineering and metalworking	79.31	131.33	184.87	259.27	321.02	345.22	335.65	316.13
Chemicals	12.86	68.61	91.93	93.61	122.79	117.82	123.23	109.35
Building materials	1.43	32.81	47.90	53.16	62.34	58.83	49.71	62.40
Wood and paper	1.91	9.19	15.04	20.08	15.03	17.06	20.05	39.13
Textile	0.00[b]	0.00	0.00	0.00	0.00	0.00	0.00	0.00
Other light industry	24.04	27.93	58.78	77.00	115.02	136.09	151.94	140.68
Food and food processing	12.98	19.49	29.61	30.22	32.44	38.47	44.23	39.53
Total industry	211.64	757.59	951.96	1151.25	1413.34	1599.01	1721.67	1877.55
Construction	63.57	87.45	121.99	154.87	238.17	269.39	277.99	277.44
Agriculture, including tractors	165.86	280.18	427.94	485.90	583.25	660.31	726.82	764.60
Transport and communication	425.59	809.12	970.72	1124.55	1196.10	1421.97	1587.01	1735.83
Trade	0.00	0.00	0.00	0.00	0.00	0.00	0.00	0.00
Other	114.47	200.98	353.78	361.45	403.76	502.87	514.96	550.36
Grand total	981.13	2135.32	2826.39	3277.72	3834.62	4453.55	4828.45	5205.78

a) Components may not add to totals because of rounding.
b) Less than 0.01 million US dollars.
Source: East-West Technology Transfer Data Base, OECD, Paris.

OECD SALES AGENTS
DÉPOSITAIRES DES PUBLICATIONS DE L'OCDE

ARGENTINA – ARGENTINE
Carlos Hirsch S.R.L., Florida 165, 4° Piso (Galería Guemes)
1333 BUENOS AIRES, Tel. 33.1787.2391 y 30.7122

AUSTRALIA – AUSTRALIE
D.A. Book (Aust.) Pty. Ltd.
11-13 Station Street (P.O. Box 163)
MITCHAM, Vic. 3132. Tel. (03) 873 4411

AUSTRIA – AUTRICHE
OECD Publications and Information Center
4 Simrockstrasse 5300 Bonn (Germany). Tel. (0228) 21.60.45
Local Agent/Agent local :
Gerold and Co., Graben 31, WIEN 1. Tel. 52.22.35

BELGIUM – BELGIQUE
Jean De Lannoy, Service Publications OCDE
avenue du Roi 202, B-1060 BRUXELLES. Tel. 02/538.51.69

CANADA
Renouf Publishing Company Limited/
Éditions Renouf Limitée Head Office/Siège social – Store/Magasin :
61, rue Sparks Street,
OTTAWA, Ontario K1P 5A6. Tel. (613)238-8985. 1-800-267-4164
Store/Magasin: 211, rue Yonge Street,
TORONTO, Ontario M5B 1M4. Tel. (416)363-3171
Regional Sales Office/
Bureau des Ventes régional :
7575 Trans-Canada Hwy., Suite 305,
SAINT-LAURENT, Québec H4T 1V6. Tél. (514)335-9274

DENMARK – DANEMARK
Munksgaard Export and Subscription Service
35, Nørre Søgade
DK 1370 KØBENHAVN K. Tel. +45.1.12.85.70

FINLAND – FINLANDE
Akateeminen Kirjakauppa
Keskuskatu 1, 00100 HELSINKI 10. Tel. 65.11.22

FRANCE
OCDE, 2, rue André-Pascal, 75775 PARIS CEDEX 16
Tel. (1) 45.24.82.00
Librairie/Bookshop : 33, rue Octave-Feuillet,
75016 PARIS. Tél. (1) 45.24.81.67 ou (1) 45.24.81.81
Principal correspondant :
13602 AIX-EN-PROVENCE : Librairie de l'Université.
Tél. 42.26.18.08

GERMANY – ALLEMAGNE
OECD Publications and Information Center
4 Simrockstrasse 5300 BONN Tel. (0228) 21.60.45

GREECE – GRÈCE
Librairie Kauffmann, 28 rue du Stade,
ATHÈNES 132. Tel. 322.21.60

HONG-KONG
Government Information Services,
Publications (Sales) Office,
Beaconsfield House, 4/F.,
Queen's Road Central

ICELAND – ISLANDE
Snaebjörn Jönsson and Co., h.f.,
Hafnarstraeti 4 and 9, P.O.B. 1131, REYKJAVIK.
Tel. 13133/14281/11936

INDIA – INDE
Oxford Book and Stationery Co. :
NEW DELHI-1, Scindia House. Tel. 45896
CALCUTTA 700016, 17 Park Street. Tel. 240832

INDONESIA – INDONÉSIE
PDIN-LIPI, P.O. Box 3065/JKT., JAKARTA, Tel. 583467

IRELAND – IRLANDE
TDC Publishers – Library Suppliers
12 North Frederick Street, DUBLIN 1 Tel. 744835-749677

ITALY – ITALIE
Libreria Commissionaria Sansoni :
Via Lamarmora 45, 50121 FIRENZE. Tel. 579751/584468
Via Bartolini 29, 20155 MILANO. Tel. 365083
Sub-depositari :
Ugo Tassi
Via A. Farnese 28, 00192 ROMA. Tel. 310590
Editrice e Libreria Herder,
Piazza Montecitorio 120, 00186 ROMA. Tel. 6794628
Agenzia Libraria Pegaso,
Via de Romita 5, 70121 BARI. Tel. 540.105/540.195
Agenzia Libraria Pegaso, Via S. Anna dei Lombardi 16, 80134 NAPOLI. Tel. 314180.
Libreria Hoepli, Via Hoepli 5, 20121 MILANO. Tel. 865446
Libreria Scientifica, Dott. Lucio de Biasio "Aeiou"
Via Meravigli 16, 20123 MILANO Tel. 807679
Libreria Zanichelli
Piazza Galvani 1/A, 40124 Bologna Tel. 237389
Libreria Lattes, Via Garibaldi 3, 10122 TORINO. Tel. 519274
La diffusione delle edizioni OCSE è inoltre assicurata dalle migliori librerie nelle città più importanti.

JAPAN – JAPON
OECD Publications and Information Center,
Landic Akasaka Bldg., 2-3-4 Akasaka,
Minato-ku, TOKYO 107 Tel. 586.2016

KOREA – CORÉE
Pan Korea Book Corporation,
P.O. Box n° 101 Kwangwhamun, SÉOUL. Tel. 72.7369

LEBANON – LIBAN
Documenta Scientifica/Redico,
Edison Building, Bliss Street, P.O. Box 5641, BEIRUT.
Tel. 354429 – 344425

MALAYSIA – MALAISIE
University of Malaya Co-operative Bookshop Ltd.
P.O. Box 1127, Jalan Pantai Baru
KUALA LUMPUR. Tel. 577701/577072

THE NETHERLANDS – PAYS-BAS
Staatsuitgeverij, Verzendboekhandel,
Chr. Plantijnstraat 1 Postbus 20014
2500 EA S-GRAVENHAGE. Tel. nr. 070.789911
Voor bestellingen: Tel. 070.789208

NEW ZEALAND – NOUVELLE-ZÉLANDE
Publications Section,
Government Printing Office Bookshops:
AUCKLAND: Retail Bookshop: 25 Rutland Street,
Mail Orders: 85 Beach Road, Private Bag C.P.O.
HAMILTON: Retail: Ward Street,
Mail Orders, P.O. Box 857
WELLINGTON: Retail: Mulgrave Street (Head Office),
Cubacade World Trade Centre
Mail Orders: Private Bag
CHRISTCHURCH: Retail: 159 Hereford Street,
Mail Orders: Private Bag
DUNEDIN: Retail: Princes Street
Mail Order: P.O. Box 1104

NORWAY – NORVÈGE
Tanum-Karl Johan a.s
P.O. Box 1177 Sentrum, 0107 OSLO 1. Tel. (02) 80.12.60

PAKISTAN
Mirza Book Agency, 65 Shahrah Quaid-E-Azam, LAHORE 3.
Tel. 66839

PORTUGAL
Livraria Portugal, Rua do Carmo 70-74,
1117 LISBOA CODEX. Tel. 360582/3

SINGAPORE – SINGAPOUR
Information Publications Pte Ltd,
Pei-Fu Industrial Building,
24 New Industrial Road N° 02-06
SINGAPORE 1953, Tel. 2831786, 2831798

SPAIN – ESPAGNE
Mundi-Prensa Libros, S.A.
Castelló 37, Apartado 1223, MADRID-28001, Tel. 431.33.99
Libreria Bosch, Ronda Universidad 11, BARCELONA 7.
Tel. 317.53.08, 317.53.58

SWEDEN – SUÈDE
AB CE Fritzes Kungl Hovbokhandel,
Box 16 356, S 103 27 STH, Regeringsgatan 12,
DS STOCKHOLM. Tel. 08/23.89.00
Subscription Agency/Abonnements:
Wennergren-Williams AB,
Box 30004, S104 25 STOCKHOLM. Tel. 08/54.12.00

SWITZERLAND – SUISSE
OECD Publications and Information Center
4 Simrockstrasse 5300 BONN (Germany). Tel. (0228) 21.60.45
Local Agents/Agents locaux
Librairie Payot, 6 rue Grenus, 1211 GENÈVE 11. Tel. 022.31.89.50

TAIWAN – FORMOSE
Good Faith Worldwide Int'l Co., Ltd.
9th floor, No. 118, Sec. 2,
Chung Hsiao E. Road. TAIPEI. Tel. 391.7396/391.7397

THAILAND – THAILANDE
Suksit Siam Co., Ltd., 1715 Rama IV Rd,
Samyan, BANGKOK 5. Tel. 2511630

TURKEY – TURQUIE
Kültur Yayinlari Is-Türk Ltd. Sti.
Atatürk Bulvari No : 191/Kat. 21
Kavaklidere/ANKARA. Tel. 17 02 66
Dolmabahce Cad. No : 29
BESIKTAS/ISTANBUL. Tel. 60 71 88

UNITED KINGDOM – ROYAUME-UNI
H.M. Stationery Office,
P.O.B. 276, LONDON SW8 5DT.
(postal orders only)
Telephone orders: (01) 622.3316, or
49 High Holborn, LONDON WC1V 6 HB (personal callers)
Branches at: EDINBURGH, BIRMINGHAM, BRISTOL, MANCHESTER, BELFAST.

UNITED STATES OF AMERICA – ÉTATS-UNIS
OECD Publications and Information Center, Suite 1207,
1750 Pennsylvania Ave., N.W. WASHINGTON, D.C.20006 – 4582
Tel. (202) 724.1857

VENEZUELA
Libreria del Este, Avda. F. Miranda 52, Edificio Galipan,
CARACAS 106. Tel. 32.23.01/33.26.04/31.58.38

YUGOSLAVIA – YOUGOSLAVIE
Jugoslovenska Knjiga, Knez Mihajlova 2, P.O.B. 36, BEOGRAD.
Tel. 621.992

Les commandes provenant de pays où l'OCDE n'a pas encore désigné de dépositaire peuvent être adressées à :
OCDE, Bureau des Publications, 2, rue André-Pascal, 75775 PARIS CEDEX 16.

Orders and inquiries from countries where sales agents have not yet been appointed may be sent to:
OECD, Publications Office, 2, rue André-Pascal, 75775 PARIS CEDEX 16.

69131-11-1985

OECD PUBLICATIONS, 2, rue André-Pascal, 75775 PARIS CEDEX 16 - No. 43427 1985
PRINTED IN FRANCE
(92 85 04 1) ISBN 92-64-12779-8